Synthesis Lectures on Engineering, Science, and Technology

The focus of this series is general topics, and applications about, and for, engineers and scientists on a wide array of applications, methods and advances. Most titles cover subjects such as professional development, education, and study skills, as well as basic introductory undergraduate material and other topics appropriate for a broader and less technical audience.

Ronald Laymon · Allan Franklin

Case Studies in Experimental Physics

Why Scientists Pursue Investigation

 Springer

Ronald Laymon
The Ohio State University
Columbus, OH, USA

Allan Franklin
University of Colorado Boulder
Boulder, CO, USA

ISSN 2690-0300   ISSN 2690-0327   (electronic)
Synthesis Lectures on Engineering, Science, and Technology
ISBN 978-3-031-12610-9   ISBN 978-3-031-12608-6   (eBook)
https://doi.org/10.1007/978-3-031-12608-6

This Springer imprint is published by the registered company Springer Nature Switzerland AG
The registered company address is: Gewerbestrasse 11, 6330 Cham, Switzerland

Acknowledgments

This book has been a collaboration, so anyone who has helped one of us has, in fact, helped both of us. With that in mind, we would like to express our gratitude to the following people. As always, members of the high-energy physics group at the University of Colorado have been available to answer questions and discuss technical issues and have provided important material. They are John Cumalat, Shanta DeAlwis, Tom DeGrand, Oliver DeWolfe, Bill Ford, Alysia Marino, Keith Ulmer, Steve Wagner, and Eric Zimmerman. Our gratitude for such support goes also to Dave Christen at the Oak Ridge National Laboratory. On a more personal note, our words are insufficient to express our gratitude for the support, encouragement and, most importantly, inspiration sent our way by Cynthia Betts and Sandy Christen.

Contents

Introduction

1

In 1977 Larry Laudan drew attention to a distinction that can profitably be made between two different sorts of evaluation of a scientific theory or research tradition. On the one hand there is what he referred to as "the context of acceptance" whereby, "scientists… choose to accept one among a group of competing theories and research traditions, i.e., *to treat it as if it were true*" (Laudan, 1977, p. 108). While on the other hand, there is "the context of pursuit" whereby, "*scientists can have good reasons for working on theories that they would not accept….*" (p. 110).

Laudan's distinction is appealing and highly suggestive precisely because it draws our attention to the fact that even if the experimental data are not sufficient to garner *acceptance* they nevertheless may be sufficient to warrant further *pursuit* of the theory at issue. Laudan's interest in the distinction was motivated by its utility in developing an account that's both historically accurate and normatively sound in cases when there's a decision to be made "to accept one *among a group of competing theories* and research traditions" (p. 108).

Having the distinction between acceptance and pursuit in hand makes it clear that acceptance need not disqualify the vanquished from further pursuit and development. Laudan gives two examples of the sort of competitive situation he has in mind. The first is that of the battle for acceptance of either "the Galilean research tradition" or "the cosmological tradition of Aristotle and Ptolemy" where Aristotle and Ptolemy had the initial lead with respect to *acceptance* but where Galileo was nevertheless worthy of *pursuit*. Similarly for Laudan's second example where "Dalton's early atomic doctrine could claim nothing like the overall problem-solving success of [the much older] elective affinity chemistry" but nevertheless Dalton was ultimately able "to predict… what are now called the laws of definite and multiple proportions." (See Laudan, 1977, pp. 112–113).

The questions immediately raised by Laudan's distinction are what are the considerations that warrant pursuit despite the existence of a historically accepted competitor, and

R. Laymon and A. Franklin, *Case Studies in Experimental Physics*,
Synthesis Lectures on Engineering, Science, and Technology,
https://doi.org/10.1007/978-3-031-12608-6_1

what warrants acceptance in the first place. With respect to acceptance Laudan's position is that "we should accept at any one time those theories or research traditions which have shown themselves to be the most successful problem solvers," where the problems to be solved divide into the "empirical" and the "conceptual." The expression "at any time" is important because it signals that acceptance may be rescinded if a competitor at a later time manages to surpass the originally accepted theory or program. The point then for allowing for pursuit is not to foreclose what may be a promising alternative from further development. And "given the fact that even our best theories may fall into crisis" it's obviously a good idea "to have alternative 'backup' theories around." (Šešelja & Straßer, 2014, pp. 3112–3113).

Putting aside for the moment the question of acceptance,[1] this leaves the question of what are the relevant factors that come into play when making such determinations of *pursuit worthiness.* Adopting for the moment Laudan's general division of empirical and conceptual problem solving an obvious initial answer is that what matters is the *promise of success* with respect to empirical and conceptual problem solving. But how is such promise to be shown? The leading candidates that have been championed in this regard fall into three general categories.

First, there are *formal considerations* such as simplicity, coherence, robustness with respect to the usual infirmities that affect theories in their initial stages, and not being ad hoc. Insofar as an up-and-coming theoretical alternative may not be fully mature in its formulation such considerations should be tempered to be understood as showing promise for being sufficiently simple, coherent, robust, and not ad hoc.

Second, there is *explanatory power* with respect to unexplained or inadequately explained experimental phenomena where those phenomena have been deemed for good reason to warrant explanation. Here we note that while a successful explanation is essential for acceptance, it is not necessary for being pursuit worthy because a promising potential for a successful explanation is enough to serve as a supporting factor for being pursuit worthy.

Third, because a decision to pursue is a decision to act, there are *practical considerations* such as expense, availability of the necessary theoretical and experimental expertise, and more generally the possibility, given current capabilities, of mounting an experimental test in the first place. After all, why commit to pursuit if it's not reasonably possible to execute.[2]

The reader may have noticed that the above factors that come into play when making a determination of pursuit worthiness are not described as being compared with an existing

---

[1] Which may after all be construed as a matter of greater pursuit worthiness than its alternatives so that it becomes a matter of prudence that some research be conducted on the less pursuit worthy alternatives.

[2] For elaborations on these categories see: Achinstein (1993), Cabera (2021), Franklin (1993), Laudan (1977), Lichtenstein (2021), Nickles (1981), Nyrup (2015), Patton (2012), and Šešelja and Straßer (2014).

accepted theory or program. They could be so relativized, but they need not. Moreover, insofar as there may be a comparative element in considerations of pursuit worthiness, it could be with respect to other theories or programs that are not accepted (as true) but are competing in terms of relative pursuit worthiness. In short, the situation could be one where there are competitors for the time and attention of scientific practitioners but where none has received acceptance. Thus, in what follows we will proceed on the assumption that questions of pursuit worthiness may have a life of their own and do not require the existence of an accepted alternative.

What though about requiring consistency with existing well-regarded theories that are not at issue? This because there must be some agreed basis for a commonality of relevant experimental testing, and more generally for some agreed standards for purposes of comparison. Such a requirement seems obvious enough. But while obvious, the question arises as to how much consistency with existing standards is required? If a great deal, then there is the risk of disqualifying pursuit options before they've had an adequate chance to shine. If only very little, then there's the risk of an unmanageably large number of options with many of questionable pedigree.[3] For the record, general problems of this sort do not arise in the case studies that we will be dealing with. This because in these cases questions of pursuit worthiness arose within rather circumscribed communities of scientific practitioners. Even so, as we shall see, serious questions were raised as to whether well-established principles such as conservation of energy and quantization at the sub-atomic level were to be maintained in the face of conflicting experimental results. Similarly, there were questions as to the durability of parity conservation and lepton conservation. How sacrosanct was such conservation to be considered with respect to the pursuit worthiness of their alternatives?

But rather than continue along the above lines of investigation into the pursuitworthiness of *theoretical* alternatives we have decided instead to tack in a different direction, namely, one motivated by the idea that questions of pursuit worthiness also apply to scientific *experiments* and *experimental programs*. Thus, an experimental result may be accepted as is, or determined to be worthy of pursuit but not acceptance, or just plain rejected. Similarly for experimental traditions or programs.[4] And because the pursuit worthiness of a theory or research program is based in large part on its *promise* of future experimental support, along with whatever existing experimental support it may have, decisions as to the pursuit worthiness of an experiment or experimental program will similarly depend on the *promise* of a proposed experiment to deliver the goods and how well experiments of the same or similar type have fared in the past. Assuming this to be so, our expectation is that a focus on the pursuit worthiness of experimentation will

---

[3] For an extended analysis of where to draw the line when it comes to specifying what existing standards are to be taken into account when deciding pursuit worthiness see Shaw (2022).

[4] While Laudan tends to speak of "research traditions" we think "research program" has a forward-looking connotation that better captures the sense of what's involved when deciding to pursue a promising theoretical or experimental development.

uncover complex and revealing relationships between the acceptance and pursuit of theories and the acceptance and pursuit of the corresponding experimental evidence. That is the guiding intuition that has motivated our approach and which we have attempted to make good on with the case studies presented here.

Our first case study deals with the early history of the discovery and further investigation of beta decay. During this time there was very little by way of theoretical prediction. It was, instead, a time of experimental exploration where theories were developed more to accommodate the experimental results than to predict them. But even though the process was in the main ad hoc and after the fact, there was no sin in that because there wasn't yet a basis on which to propose more substantive theories.

A recurring experimental problem was whether an apparatus and procedure were worthy of continued pursuit despite of difficulties with controlling confounding effects. Thus, it was a question of determining what the prospects were for getting such confounding factors under control. In particular, there were two such problems that impacted resolution when it came to isolating for investigation a bundle of β-rays of restricted diversity. Because the apertures used to effectuate such isolation were necessarily of finite size, there were unavoidable limitations on resolution. So, the question was what the pursuit worthy options for experimental refinement were that would minimize such loss of resolution. More difficult to deal with as a confounding factor was the scatter of the β-particles as they bounced about in resistance to various schemes of the experimenters to filter out such unruly conduct. As we'll see, an experimental approach with a long history of qualified success came to outlive its usefulness because of the persistence of such scatter.

There also came a point in the history of beta decay when two competing experimental approaches (the "electric" and the "photographic") led to what appeared to be irreconcilable differences. So, the problem was to come up with pursuit worthy avenues of apparatus improvement that would allow for either reconciliation or outright rejection of one of the experimental programs. An essential component of efforts to deal with such problems was the replication of earlier experimental results where the replications incorporated pursuit worthy improvements that promised to solve or mitigate the problems with experimental accuracy and reliability.

In 1927 agreed upon experimental results came into a striking conflict with the hitherto agreed upon set of principles: (1) conservation of energy and (2) quantization at the atomic and sub-atomic levels. And here experimentation won out so that the only pursuit worthy option was that one or more members of the combination had to be rejected. Which in turn meant that members of the physics community had to make an individual choice as to what was pursuit worthy as a matter of theoretical and associated experimental development.

Our other cases studies deal with more contemporary examples where existing theory played a more pronounced role in predicting and explaining experimental results. These cases will illustrate the importance of the availability of feasible apparatus and experimental expertise. Also, in the mix of considerations of pursuit worthiness was the importance of the theoretical issues involved such as parity and lepton conservation.

At a more particular level, we show how, in the case of the Wu experiment, the experimental result was accepted (with some assistance from two other related experiments) as demonstrating that parity was not conserved in weak force interactions. But that same experimental result was only deemed pursuit worthy when it came to a more definitive determination of the asymmetry coefficient. Another of our case studies, that dealing with the Fifth Force, shows that experiments may sometimes be deemed not pursuit worthy even though there is no plausible explanation as to why and how they have failed. Here the argument for excommunication was that the experiment in question should be deemed a failure because other experiments designed with the same purpose led to conflicting results. The evidence against that result was overwhelming.

Some of our examples deal with the pursuit worthiness of further investigation of the statistical analysis used to justify the claim of an experimental result. And here we note that in high-energy physics as well as in gravity-wave experimentation, there is a formal criterion for the acceptance of a claimed discovery, namely, that the observed signal must be five standard deviations (5σ) above background. Thus, such an experimental result should be accepted as a genuine discovery only when this criterion is satisfied. Assuming, however, that the statistical analysis given in support is itself accepted, and not deemed only worthy of further pursuit. We will review a striking example (from research on the existence of neutrinoless double beta decay) where exactly this question came to the fore. Another example (dealing with the response to the muon g-2 experiment) deals with a situation where further experimental and theoretical investigation were both deemed pursuit worthy.

This concludes our introductory remarks. It's now time to get on with the specifics.

References

Achinstein, P. (1993). How to defend a theory without testing it: Niels Bohr and the logic of pursuit. *Midwest Studies in Philosophy, 18*, 90–120.

Cabera, F. (2021). String Theory, non-empirical theory assessment, and the context of pursuit. *Synthese, 198*(Suppl 16), S3671-3699.

Franklin, A. (1993). Discovery, pursuit, and justification. *Perspectives on Science, 1*, 252–284.

Laudan, L. (1977). *Progress and its problems: Toward a theory of scientific growth*. University of California Press.

Lichtenstein, E. I. (2021). (Mis)Understanding scientific disagreement: Success versus pursuit-worthiness in theory choice. *Studies in History and Philosophy of Science, 85*, 166–175.

Nickles, T. (1981). What is a problem that we may solve it? *Synthese, 47*, 85–118.

Nyrup, R. (2015). How explanatory reasoning justifies pursuit: A Peircean view of IBE. *Philosophy of Science, 82*, 749–760.

Patton, L. (2012). Experiment and theory building. *Synthese, 184*, 235–246.

Šešelja, D., & Straßer, D. (2014). Epistemic justification in the context of pursuit: A coherentist approach. *Synthese, 191*, 3111–3141.

Shaw, J. (2022). On the very idea of pursuitworthiness. *Studies in History and Philosophy of Science, 91*, 103–112

The Beta Decay Spectrum: Experimental Discovery and Pursuit

2

2.1   Discovery and Identification of α, β and γ Radiation

Our story begins in 1896 with the almost accidental discovery of radioactivity by Henri Becquerel, and Pierre and Marie Curie's discovery in 1898 of two new and more intense radioactive sources, polonium and radium.[1] The first step in deciphering the nature of the radiation emitted by uranium, called Becquerel rays, was taken by Ernest Rutherford who conducted a series of experiments on the absorption of that radiation. Using aluminum foils and measuring penetration he reported:

> It will be observed that for the first three layers of aluminum foil, the intensity of the radiation falls off according to the ordinary absorption law, and that, after the fourth thickness the intensity of the radiation is only slightly diminished by adding another eight layers (Rutherford, 1899, p. 115).[2]

On the basis of this difference in absorption, Rutherford concluded:

> These experiments show that the uranium radiation is complex, and that there are present at least two distinct types of radiation – one that is vary readily absorbed, which will be termed for convenience the α radiation, and the other of a more penetrative character, which will be termed the β radiation (Rutherford, 1899, p. 116).

---

[1] Because our intent here is entirely introductory, for details and references we refer the reader to Rutherford (1904, pp. 90–148), Rutherford (1913, pp. 114–295), Kohlrausch (1928, pp. 189–203, 414–421), Jensen (2000, pp. 1–8), and Franklin and Marino (2020, pp. 1–19).

[2] By the "ordinary absorption law" Rutherford meant that the intensity of the radiation "diminishes in a geometrical progression with the thickness of metal... i.e. according to an ordinary absorption law." See Rutherford (1899, p. 115).

© The Author(s), under exclusive license to Springer Nature Switzerland AG 2022   7
R. Laymon and A. Franklin, *Case Studies in Experimental Physics*,
Synthesis Lectures on Engineering, Science, and Technology,
https://doi.org/10.1007/978-3-031-12608-6_2

In sum, the first four foils each considerably reduced, and finally eliminated, the α radiation. The remaining β radiation was then only slightly reduced by each of the following foils.

The beta rays were not long thereafter identified as electrons. By contrast, it was believed initially that the α particles were electrically neutral because they could not be deflected by a magnetic field. Rutherford found, however, that they could be deflected in the same direction as a positive charge when he applied a strong magnetic field. Subsequent work by Rutherford and others showed that the α particles were helium ions. Rutherford later used the scattering of these high-energy α particles from gold and other foils to argue for the nuclear model of the atom—a very small, heavy, positively charged nucleus surrounded by negatively charged electrons, a miniature solar system. On the basis of Rutherford's nuclear model of the atom, the α particles were considered to be helium nuclei. Finally, to complete what turned out to be a triumvirate, γ rays were discovered by Paul Villard in 1900, and were found to be high-energy electromagnetic radiation that was electrically neutral.

To complete our brief introduction of α, β and γ radiation we need to highlight an important further distinction between the α and β particles. As William Bragg was able to show, α particles were emitted with the same energy in each particular radioactive decay and had a constant range when moving through matter. β particles, by contrast, were thought to be emitted over an extended range of velocities and moreover had a decreasing range when moving through matter. With regard to the behavior of a beam of β particles (i.e., electrons) in air, Bragg noted that such a beam would become diffuse because of the scattering of the electrons and that the electrons would lose energy due to ionization.

> If such a jet of electrons be projected into the air, some will go far without serious encounter with the electrons of air molecules; some will be deflected at an early date from their original directions. The general effect will be that of a stream whose borders become ill-defined, which weakens as it goes, and is surrounded by a haze of scattered electrons. At a certain distance from the source all definition is gone, and the force of the stream is spent. There is a second cause of the gradual 'absorption' of a stream of β rays. Occasionally an electron in passing through an atom goes so near to one of the electrons of the atom as to tear it from its place, and so to cause ionization. In doing so, it expends some of its energy (Bragg, 1904, p. 719).

The difference between the behavior of the α and β particles was due to the difference in their interactions with matter. Alpha particles lose energy almost solely by ionization, whereas electrons lose energy by several processes including both ionization and by scattering, and by other processes unknown to physicists in the early twentieth century. It was believed, at the time, that β particles would follow an exponential absorption law when they passed through matter. This was a reasonable assumption. If electron absorption was dominated by the scattering of electrons out of the beam, and if the scattering probability per unit length was constant, then this leads to an exponential absorption law. As Bragg stated:

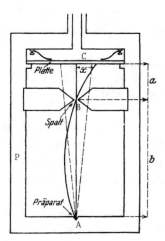

Fig. 2.1 Becquerel's apparatus as later refined and used by Hahn and Meitner. *Source* Kohlrausch (1928)

[I]t is clear that β rays are liable to deflexion through close encounters with the electrons of atoms; and therefore the distance to which any given electron is likely to penetrate before it encounters a serious deflexion *is a matter of chance*. This, of course, brings in an exponential law (p. 720, emphasis added).

2.2 Early Experiments and the Velocity Distribution of β-Rays

Because our focus in this chapter will be with β-rays, and in particular the experimental discovery and subsequent explanation of what became known as the *beta spectrum*, we 'll move on to a brief review of three early experiments that dealt with the velocity distribution of β-rays. Becquerel, in particular, was especially creative and efficient in this regard and in 1900 devised two rather different experimental approaches.[3]

In the first he used an apparatus that was essentially identical to that which was still in use over a decade later. See Fig. 2.1 which depicts Becquerel's apparatus in one of its later reincarnated forms.[4]

The importance of this particular experimental design is attested to by the fact that it made starring appearances in Rutherford (1904, pp. 98–100) as well as in its sequel (Rutherford, 1913, pp. 196–197). In essence, the radioactive source was placed at the bottom of a lead box, a magnetic field was applied, and the β-rays made their way (either through a narrow cylinder or in later variants a sequence of aperture slits) to a photograph

---

[3] See Becquerel (1900a, 1900b, 1900c, 1901).

[4] For confirmation on this continuity of design see Kohlrausch (1928, pp. 202–203).

plate located at the top of the apparatus. The result was that "each part of the plate is acted on by rays of a definite curvature" (Rutherford, 1904, p. 98, 1913, p. 196).

> [Thus] a diffuse impression is observed on the plate, giving, so to speak, a *continuous spectrum* of the rays and showing that the radiation is composed of rays widely different velocities (1904, p. 98, 1913, p. 196, emphasis added).

Becquerel's apparatus was also used to demonstrate that "the most deviable rays are those most readily absorbed by matter." This was accomplished by "placing screens of various thickness on the [photographic] plate" and noting the correspondences between distance of the source from the plate and the thickness of the absorbing screens (1904, pp. 99–100, 1913, p. 197).

Becquerel also devised a more sophisticated apparatus that promised a "significant improvement in the quantitative determination" of the variations in β-particle velocity (Kohlrausch, 1928, p. 198). The apparatus is depicted in Fig. 2.2 where the radium beta source was located in a lead groove beneath the slit *Sp* which led into two eccentrically placed cylinders, each with slits on one side.[5] The emitted β-particles were then deflected by a magnetic field which caused them to move through the apparatus in circular orbits where the radius of the orbits is proportional to their velocity. The function of the two cylinders and the corresponding slits was to impose an improved degree of collimation to the emitted β-particles. In addition, by rotating the cylinders the entire range of velocity values could be extracted because the lower velocity electrons had a larger deflection and a smaller radius of curvature, while the higher velocity electrons had smaller deflection and thus a larger radius of curvature. Detection of the escaping β-particles was by means of a photographic plate that was mounted parallel to the plane of the circles. Despite the promise of a "significant improvement in the quantitative determination" the results were disappointing and delivering on the promise had to wait till further development of experimental apparatus.[6] In any case, Rutherford's appraisal was that:

> The deviable rays from radium are complex, i.e. they are composed of a flight of particles projected with a *wide range of velocity*. … The complexity of the radiation has been shown very clearly by Becquerel …. (Rutherford, 1904, p. 98, 1913, p. 196, emphasis added)

In 1904, Friedrich Paschen published the results of his investigation of the velocity spectrum of beta rays from radium (Paschen, 1904). Here the aim was to make a determination of the correlation between a velocity value extracted by a varying magnetic field and the number of β-particles that were associated with that velocity.

As shown in Fig. 2.3, the RaB source was sealed in a small glass vessel in the center (at b) of six lead blades which were encased by an insulated lead cylinder which was

---

[5] The origin of this figure is uncertain because none of the references to Becquerel cited at Kohlrausch (1928, p. 198) contains this figure.

[6] See Jensen (2000, p. 4).

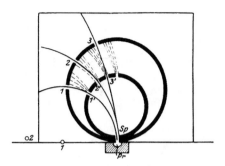

Fig. 2.2 Becquerel's apparatus that promised a "significant improvement in the quantitative determination" of β-particle velocity. *Source* Kohlrausch (1928)

connected to an electroscope (at A).[7] As in the case of Becquerel's experiment, a magnetic field (of varying intensities) was imposed which caused the β-particles to be deflected into circular orbits. As summarized by Rutherford, Paschen measured "the total charge carried by the β rays which were not bent away by a known magnetic field" (Rutherford, 1913, p. 251), as shown on curve I in Fig. 2.4. Paschen was then able to determine the total remaining charge as a function of HR/2 (the product of the field strength and the radius of curvature divided by 2), where the first derivative of this curve gives a value of the number of β-particles as a function of their velocity, as shown by curve II on Fig. 2.4.[8]

2.3   Schmidt and the Absorption and Velocity of β-Rays

In addition to the attempts by Becquerel and Paschen to experimentally determine the general nature of the beta spectrum, there were experimental efforts that were more specifically directed at determining whether β-particle absorption followed an exponential or some other law-like regularity. By and large, but not always, these experimental investigations yielded an exponential result if only approximately.[9] We'll pass over those efforts and move directly to the experiments and results in this regard by Heinrich Schmidt since they were most directly relevant in terms of the pursuit worthiness of the later

---

[7] This figure does not appear in Paschen (1904) and was apparently separately prepared for publication in Kohlrausch (1928, p. 200).

[8] For more detailed summaries of Paschen's measurement and analysis (including his assumption that charge deposited by β-particles of velocity v is a function of the number of β-particles with that velocity) see Kohlrausch (1928, pp. 199–201) and Jensen (2000, p. 5).

[9] For brief reviews of these experiments see Jensen (2000, pp. 7–8), Rutherford (1904, pp. 112–115) and Rutherford (1913, pp. 223–226).

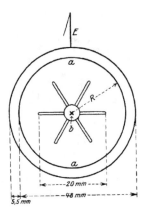

Fig. 2.3 Paschen's apparatus that was used to investigate the velocity spectrum of beta rays. *Source* Kohlrausch (1928)

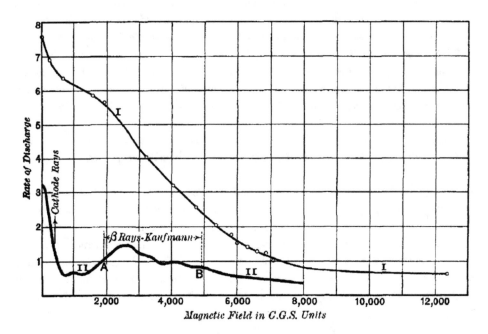

Fig. 2.4 Paschen's measured values where curve I is the total charge of the β-rays that remained after application of a magnetic field, and curve II is the first derivative which is a measure of the number of particles as a function of their velocity. *Source* Paschen (1904)

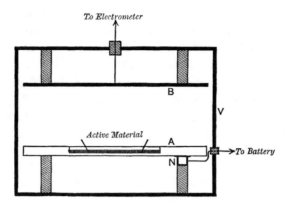

Fig. 2.5 The type of apparatus typically used, including by Schmidt, for testing the absorption of β-rays. *Source* Rutherford (1913)

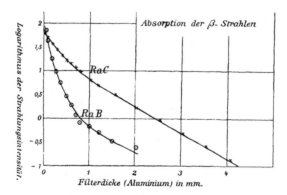

Fig. 2.6 Schmidt's absorption curves where the logarithm of the β-ray intensity values is plotted against filter thickness, and where each curve was calculated as the a sum of exponential functions. *Source* Schmidt (1906)

experiments on which we'll focus our attention. Schmidt's experimental apparatus was essentially similar to that shown in Fig. 2.5.[10]

Figure 2.6 shows the absorption curves that Schmidt obtained for β-particles emitted from radium B and from radium C.

For absorption thickness greater than 1 mm the (logarithmic) curve for RaC is a straight line while that of RaB is close to being a straight line. Thus, as Schmidt realized, it follows

---

[10] While Schmidt's publication does not include an illustration of his apparatus, his description at Schmidt (1906, p. 764) is consistent with Fig. 2.5 which occurs at Rutherford (1904, p. 82) and again at Rutherford (1913, p. 106) where the apparatus is described as "very convenient for many measurements in radio-activity."

that "within certain filter thicknesses the points of the curve absorb according to a pure exponential function" (Schmidt, 1906, p. 765).[11] But, as should be emphasized, this "pure exponential function" only held for the specified range. In response, Schmidt suggested the following explanation of the *complete* data set.

Couldn't this also be explained by saying that among all the beta rays there exists a certain group with constant absorption coefficient? Indeed, couldn't we even go one step further and explain the total effect of the beta rays [of a single substance] by assuming that a few beta-ray groups have a constant absorption coefficient? It follows that the radiation intensity I could be expressed according to the formula

$$I = a_1 e^{-\lambda_1 d} + a_2 e^{-\lambda_2 d} + \cdots$$

where d is the filter thickness, e the basis of the natural logarithms, and a and λ certain constants (p. 765).

Applying his proposal to the RaC and RaB curves at hand, he determined that a curve (shown by the solid line in Fig. 2.6) with a very close fit for the *entire* RaC data could be obtained on the assumption that:

$$I = 1100e^{-890d} + 88e^{-80d} + 2.5e^{-13.1d}$$

Similarly, for RaB, where the assumed combination of exponentials was:

$$I = +49e^{-53d} + 25e^{-13.1d}$$

Emboldened by his success, Schmidt then drew the conclusion that:

Since the penetrating power of the beta rays can only depend on their velocity, it follows from the constancy of the absorption coefficient that when passing through matter the particles do not change their velocity at all (p. 766).

In addition, Schmidt thought it likely that a "scattering and a complete annihilation of the individual particles take place simultaneously" (p. 766). In this way changes in ionization could be accounted for as a reduction in the number of β-particles with their characteristic and unchanging velocity.

Schmidt's proposed conclusion was obviously *pursuit worthy* given its initial plausibility and importance for understanding the nature of β-rays. And as we shall see, Schmidt was not the only one who thought so. In any case, he decided to devise an experiment that would test whether β-particles did not in fact change their velocity as they passed through absorbing media.

The apparatus Schmidt employed to effectuate his experimental test is shown in Fig. 2.7 where two variations are depicted. The one on the left-hand side employed a

---

[11] Translation by Jensen (2000, p. 8).

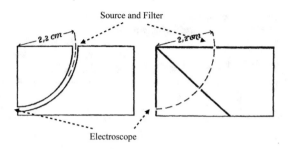

Fig. 2.7 Schmidt's experimental apparatus. *Source* Schmidt (1907), augmented by authors

block of lead with a small circular channel, while that on the right achieved the same collimating effect but with three brass diaphragms soldered on a common sheet. The RaE source was placed at the top, the entrance to the collimating channel, and a device to measure ionization placed at the exit. Using RaE was a good choice because it emitted relatively few confounding gamma rays.

Beta rays of different velocities were then selected by varying a magnetic field oriented perpendicularly to the apparatus.[12] After taking such measurements with the lead and brass versions of the apparatus, aluminum absorbing filters of different thickness were then interposed between the source and the entrance to the collimating channel. Schmidt presented his results in the graphical form shown in Fig. 2.8. By way of explanation and translation, we note that the top two curves deal respectively with the relationship between field strength (i.e., velocity) and ionization for the lead and brass versions of the apparatus when no absorbing filters were interposed. The next two curves present that relationship when aluminum filters of different thicknesses were interposed where, however, the scale of the bottom curve was magnified by a factor of ten.

There are two noteworthy features of Schmidt's data curves. First, they indicate quite clearly that while the energy of the emitted β-rays increases as the field strength (and thus the velocity of the β-particles increases) it eventually reaches a point where that energy decreases as shown by the bell-shaped curve. Second, the curves remain similar in shape (though not in amplitude) when the absorbing filters were employed, and where, most importantly, the maximums all coincide at the same magnetic field strength. From this similarity in shape and coincidence of maximums, Schmidt went on to conclude:

> [T]he energy given up by the beta rays when passing through matter does not correspond to a reduction of the particles' velocity. In all probability, an absorption, i.e., an annihilation of radiation energy, originates from the fact that for equally thick layers the same percentage of

---

[12] For a detailed explanation of how such a magnetic field works to segregate velocities see Rutherford (1913, pp. 72–75).

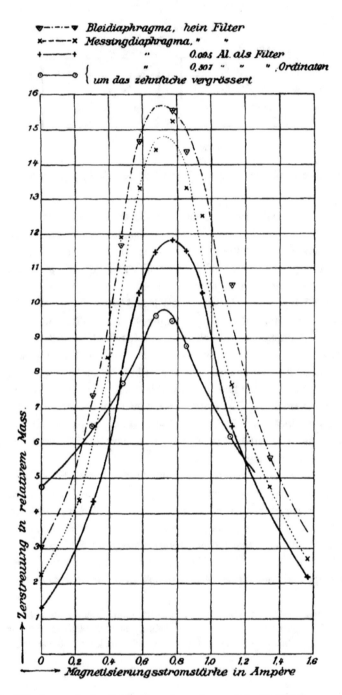

Fig. 2.8 Schmidt's β-ray velocity-distribution curves. *Source* Schmidt (1907)

particles are stopped completely each time, whereas the remaining particles fly through with undiminished velocity (Schmidt, 1907, p. 372).[13]

While Schmidt's conclusion did not gain general acceptance, it and his experimental support was most certainly pursuit worthy, and in fact, as we shall see, it became the model for later developments. And with good reason because it effectively distilled and employed the best elements of the earlier apparatus of Becquerel and Paschen. It was thus a Kuhnian paradigm of experimentation, a great success but not without its problems and with definite possibilities for improvement.

There was, however, a blemish in Schmidt's experiment. As Schmidt made clear his data did not support a pure exponential law of absorption for RaE and were not otherwise explainable in terms of a sum of exponential functions. But all was not lost because of the likely confounding effect of scattering.

Now it is quite possible that a pure exponential law holds for absorption proper, i.e., for the annihilation of radiation energy, and that the deviations from this law are only apparent. The deviations are caused precisely by the fact that part of the scattered radiation energy is not measured with the experimental arrangement chosen, and furthermore that scattered rays cover a longer distance in the filter than those passing straight through (Schmidt, 1907, pp. 370–371).

We draw this to the reader's attention because it played a major role in a later explanation as to why the coincidence of observed maximums did not support Schmidt's conclusion that β-ray velocity did not diminish on passing through absorbing materials. Thus, this apparently minor blemish would provide a telling opportunity for criticism.[14]

But before following that thread of our story, we'll briefly review another line of investigation and development where the exponential absorption of beta particles was used as a central component of a method of testing and discovering pure forms of radioactive substances. The success of the method in turn provided support for the acceptance of the hypothesis that absorption proceeded exponentially.

---

[13] Translation at Jensen (2000, p. 11).

[14] We've already discussed Bragg's warning about the effect of scattering, but for more on the then contemporary understanding see Rutherford (1913, pp. 212–222) who pointedly noted that: "It would be very convenient to call the rays which are turned back on the side of incidence the *reflected* rays, and those which pass through the plate the *transmitted* rays. In using the term 'reflected,' however, it must be remembered that the phenomenon has no analogy with light, but that the reflected rays consist of the primary β particles which have been so scattered by the atoms of matter that they emerge again in all directions" (p. 213).

2.4   Hahn and Meitner: Exponential Absorption and Homogeneity of Substance

The first of the β-ray experiments that we'll examine in detail was performed by William Wilson around 1909 and involved an extended investigation of (1) the absorption of β-rays as they passed through absorbing media, and (2) whether β-rays lost velocity or retained their velocity, as claimed by Schmidt, as they passed through such media. But to fully understand Wilson's experiment it is necessary to know something about why it was that he targeted certain additional hypotheses for experimental investigation. In other words, what made his experimental efforts especially pursuit worthy. What was at issue was a set of hypotheses about absorption and the related evidential inferences made by Otto Hahn and Lise Meitner.

Hahn and Meitner discovered that when β-rays were emitted from thorium and several other substances, their absorption when passing through media of various thickness was essentially exponential.[15]

> The fact that under our controlled conditions the absorption measurements gave a pure exponential law also for infinitely thin layers, *led us naturally to the assumption* that also the other beta-ray emitting substances, as long as they are uniform [*einheitlich*], would show a similar behavior, and that in cases where the apparent absorption coefficient decreases with increasing thickness of the penetrated layer, we were not dealing with uniform [*einheitliche*] beta-ray emitting substances (Hahn & Meitner, 1908a, p. 328).[16]

In this context the sense of *einheitlich* is that such a substance is chemically pure.[17] So the sense here is that if you've got a substance that's pure, absorption will be found to proceed exponentially. To this was added the following:

> When one considers the *analogy* with alpha rays, our [additional] assumption that uniform (*einheitliche*) beta-ray emitting products also emit *only one group of beta rays*, seems a priori more probable [than the assumption of more than one] (p. 331, emphasis added).

Keeping the analogy with alpha particles in mind, what Hahn and Meitner meant by "only one group of beta rays" is that β-rays emitted by a radioactive substance are *all emitted with the same velocity* where that velocity depends on the particular substance.[18] Consequently any experimental data that indicated a more or less continuous spectrum of

---

[15] We'll review the mathematical description of linear and exponential absorption, and absorption coefficients, when we get to Wilson's 1909 experiment.

[16] Translation by Jensen (2000, p. 13).

[17] See Rutherford (1913, pp. 568–586) and Chadwick (1921, 600–604) for brief reviews of the then contemporary understanding of the chemical properties of radioactive substances.

[18] This feature mirrors what was an accepted feature of alpha particles.

β-rays velocities was an artifact of the experimental apparatus, that is, due to secondary causes, most notably scattering.[19]

The attraction of this position is that it provides an elegantly simple account of the underlying nature of β-rays and relegates apparently contrary experimental evidence as something to be explained in terms of the particularities of the experimental apparatus. It does involve, however, rejecting Schmidt's proposal that the total effect of the beta rays of a single substance could be understood as being due to a combination of a few beta-ray groups each of which had its own absorption coefficient. Hahn and Meitner did, however, adopt Schmidt's hypothesis that when passing through matter β-rays retained their velocity but suffered a reduction in the number of particles with that velocity.

This combination of assumptions, identified by Hahn and Meitner as a "working hypothesis," drew substantial confirmation from a series of additional investigations of radioactive substances which all instantiated the working hypothesis.

> Our assumption, already made earlier, that uniform substances also emit uniform beta rays, and that their absorption in aluminum follows an exponential law, has fully retained its status *as a working hypothesis* also for actinium, and *has led to the discovery of the new groups of beta rays* (Hahn & Meitner, 1908b, p. 702, emphasis added).[20]

So, for example, when mesothorium and "radioactinium" were tested they were found not to be exponential with respect to absorption which suggested according to the working hypothesis that they were not "uniform" substances which was later confirmed by chemical means (Hahn, 1909). Similarly for radium (Hahn & Meitner, 1909a). In sum, no counter instances were discovered to the "working hypothesis." But as we shall see this comfortable coherence between working hypothesis and experimental instances was soon to be upended by William Wilson.

But before proceeding to Wilson's experiment, we need to note that in all their work Hahn and Meitner were handicapped by *not* having available the experimental means necessary to impose a varying field strength on a beam of β-particles such that those particles could be separated out according to their velocity. Instead, they had to rely on more primitive β-ray collection methods that simply showered the absorption screens with *all* of the emitted radiation mediated only by physical barriers.[21]

---

[19] This meant that Schmidt's ionization-field strength curve (when no absorbing material was involved) had to be seen as a secondary phenomenon and not as indicative of an underlying β-ray spectrum.

[20] Translation by Jensen (2000, p. 16).

[21] For a description of their apparatus see Hahn and Meitner (1908a, p. 322). For a diagram of this general type of device see Rutherford (1904, p. 82). See also Meitner's later account of the difficulties in being so experimentally handicapped (Meitner, 1964, pp. 5–6).

2.5   Wilson (1909): Absorption is Linear and Velocity Decreases

Rutherford evidently found the work of Schmidt and that of Hahn and Meitner pursuit worthy and so suggested to his student, William Wilson, that he undertake the investigation of "the connection between the absorption and velocity of the β-rays" (Wilson, 1909, pp. 612, 628). In the first of his published reports on this subject, Wilson began by stating what his experimental results would show to be false.

> It has been generally assumed that a beam of homogeneous rays is absorbed according to an exponential law, and the fact that this law holds for the rays from uranium X, actinium, and radium E has been taken as a criterion of their homogeneity (p. 612).[22]

To avoid later confusion, we note that the first conjunct is most naturally understood as a claim about experimental methodology. Something like it's been assumed that if a reliable means has been used to extract from a radioactive source a beam of β-rays that is believed to be homogeneous (in the sense that all of the rays are of the same velocity), then those rays will be absorbed according to an exponential law. On this reading, what Wilson intends to show is that he has *a more reliable means* of extracting a homogeneous beam of β-rays and that beams extracted this way are *not* absorbed according to an exponential law but rather, as we shall see, *according to a linear law*. And from this experimental result it may or may not be a short step to the ontological claim that it is a fundamental property of β-rays that they are absorbed according to a law that is linear and this experimentally discovered behavior is not just an artifact of the experimental means used.

While the second conjunct bears some resemblance to the "working hypothesis" of Hahn and Meitner, their working hypothesis is better stated as: If a well collimated beam of β-rays from a radioactive substance is found to be absorbed according to an exponential law then that substance is likely to be homogeneous in the sense of being a *chemically pure substance*, where the support for this working hypothesis is that impure substances have been found not to emit β-rays that are absorbed according to an exponential law.

Regarding the first conjunct, Wilson went on to state with more specificity that:

> Without entering at present into further details, it can be stated that the ionization [of radium] did not vary exponentially with the thickness of matter traversed, but, except for a small portion at the end of the curve, followed approximately a linear law (Wilson, 1909, p. 613).

Before proceeding further, we wish to emphasize that Wilson's response to what was an unruly situation was decidedly original and insightful. His approach was based on the anticipation that Schmidt's experimental apparatus could be refined to achieve better collimation and immunity against confounding effects. If so, then a *bundle of beta rays* could

---

[22] Wilson cites (Hahn & Meitner, 1908a, p. 321) as an example of those who have assumed this.

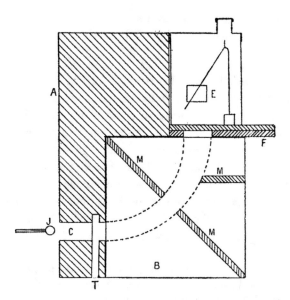

Fig. 2.9 Wilson's experimental arrangement where the absorbing material was placed immediately below the electroscope at E. *Source* Wilson (1909)

be extracted by means of a varying magnetic field where that bundle at least "approximately" consisted of parallel rays of the same speed. And that once so extracted and isolated such bundles of rays could be tested for how they were absorbed. As we shall see, Wilson's efforts paid off because his experimental results indicated that absorption for homogeneous β-rays was linear and clearly not exponential. Wilson's results thus provided the justification for a rather *different sort of pursuit worthy experimental program*: the investigation of the properties of homogeneous rays so extracted.

   Wilson's experimental apparatus is shown in Fig. 2.9.

   A radium bromide source was placed at C. The collimated β-rays from the decay of radium were bent in a circular path by a magnetic field[23] perpendicular to the plane of the paper, passed first through slits MM and F, then passed through aluminum absorbers of different thickness (placed just below the electroscope), and were detected by the ionization produced in an electroscope at E. The radius of the circular path is proportional to the velocity of the electrons. Varying the strength of the magnetic field changed the velocity of the selected electrons so that the absorption of the electrons as a function of velocity could be determined.

   Wilson stated his initial results as follows:

   [T]he ionization in the electroscope was found not to fall off, according to an exponential law, with the thickness of matter traversed, but more rapidly the greater the distance penetrated.

---

[23] Wilson notes that "[t]he field was found to be practically uniform" (p. 614).

For the rays of higher velocity, the relation was linear, except when the radiation had been cut down to a large extent, when it fell off more slowly with increasing thicknesses of matter (p. 615).

These results were graphically presented as shown in Fig. 2.10, where the upper graph shows the ionization for four different initial velocities (as determined by a variation of the magnetic field strength) as a function of absorber thickness. It is clearly linear, and not exponential. This is made clear in the lower graph in which the logarithm of the ionization of one of the curves from the upper graph is plotted against absorber thickness.

Having now determined the relationship between ionization and magnetic field strength Wilson went on to determine more formally and precisely the variation of absorption with velocity. The result of all this would provide the basis for an argument that the velocity of β-rays was reduced by passage through an absorbing media. The first step was to specify what constitutes the "linear law" that controls absorption of homogeneous β-rays.

The [linear] relation between ionization and thickness of matter traversed is given by $I = k (a - x)$, where a is the thickness of matter for which the ionization would become zero if the law were rigorously true, and ka the initial ionization [and where x is the thickness of the absorbing medium] (617).

Alternatively stated, "a is the distance from the origin of the point where the straight line if produced cuts the axis of the abscissa" (Rutherford, 1913, p. 237). See Fig. 2.10 where we have extended one of Wilson's experimentally determined linear curves to the intersection point at a. The values for the variation of a with velocity (for aluminum) were given in Table 2.1, where a was determined as described above, R was the radius of curvature as determined by the geometry of Wilson apparatus; HR the product of the magnetic field strength and the radius; and V the velocity of the β-rays was calculated by means of the formula $HR = mv/e$.[24]

The last column in Table 2.1 gives the values for λ, *the absorption coefficient,* which is indirectly defined as:

$$dI = -\lambda I \, dx$$

The role of λ therefore is to specify the rate of change of ionization with respect to a change in the thickness of the absorbing substance. Now if the process is linear, i.e., defined by $I = k (a - x)$, then we can solve for λ as follows:

$$\lambda = -(I \, dI / I \, dx) = 1/(a - x)$$

Thus, λ "increases as the thickness of matter traversed by the rays is increased." If, however, the absorption process is exponential, λ would be a constant.

---

[24] Actually, this calculation is somewhat more complicated because it requires the relativistic correction for the increase in mass at very high speeds. See Wilson (1909, pp. 617–618).

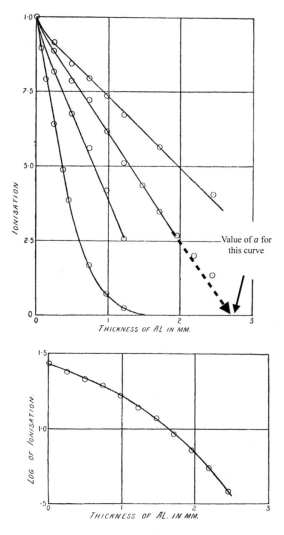

Fig. 2.10 Wilson's absorption results where the upper graph shows the ionization for four different initial velocities as a function of absorber thickness, and where the lower graph plots the logarithm of the ionization against absorber thickness. *Source* Wilson (1909), augmented by the authors showing the value of *a* for one of the curves

At this point, Wilson decided that "[b]efore entering into a discussion as to the meaning of the absorption curves obtained, it is preferable to try to *explain* why various observers have found that the rays from uranium X, radium E, and actinium are absorbed according to an exponential law with the thickness of matter traversed" (p. 621, emphasis added). Wilson's explanation is that these observers had failed to extract homogeneous beams of B-rays from their test sources.

Table 2.1 The values of a, the absorber thickness at which the ionization is zero, along with the absorption coefficient λ as a function of electron velocity

R (cm)	H.R (Gauss cm)	V (10^{10} cm)	A (mm)	Λ (cm^{-1})
4.0	1310	1.79	0.14	71.5
	1860	2.20	0.26	38.5
	2760	2.545	0.70	14.3
	3860	2.74	1.32	7.6
	4450	2.795	1.67	6.0
	5390	2.855	2.08	4.82
	6350	2.899	2.64	3.79
	8000	2.920	3.29	3.04
	8580	2.930	4.09	2.44
	8960	2.937	4.69	2.13
2.1	980	1.46	0.095	105.0
	1820	2.19	0.235	42.5
	2160	2.35	0.35	28.6
	2640	2.515	0.535	18.7
	3300	2.665	0.96	10.4
	4300	2.782	1.28	8.35
	5190	2.844	2.00	5.00
	5490	2.859	2.12	4.17

Source Wilson (1909, p. 618)

> The fact [as we have found] that homogeneous rays are not absorbed according to an exponential law *suggests* that the rays from these substances [used by these various observers] are *heterogeneous* (p. 621, italics added).

Now there is more to Wilson's "explanation" than just the claim that there was something amiss in the experimental means used by the "various observers" in question. Wilson's explanation rested on his mathematical (not experimental) proof (see pp. 622–624) of the following proposition:

> [I]t is possible to obtain a heterogeneous beam of particles, of which the different types of rays are absorbed according to a linear law, but the absorption of the whole beam takes place according to an exact exponential law (p. 624).

In other words, what Wilson is proposing by way of explanation was a *specific way* in which the various observers had been led astray, namely, that while they may have been successful in culling a bundle of rays that as a composite were absorbed exponentially, those bundles were not homogeneous (with respect to velocity) but nevertheless

had the right mixture of rays such that the resulting absorption of their composite bundles was exponential.[25] We note, however, that Wilson did not go on to explain what
had gone wrong in these experiments that led to the inhomogeneity of their β-ray beams,
whether, for example there was a failure of collimation or an unsuspected impurity in the
radioactive sources used.

Having completed his explanation of where his scientific contemporaries had gone
wrong Wilson turned to a consideration of the "meaning" of his absorption data. And
here the issue was the relevance of that data for deciding between the following two ways
in which the absorption of a beam of particles can take place.

> In one the [individual] particles *lose energy* as they pass through matter and finally cease to be
> effective as ionizing agents. ... In the other the particles are stopped in mid-career while their
> velocity is still high, and [where] particles are assumed to pass through matter with a constant
> velocity, a sudden stopping of a certain proportion of the rays taking place in each thin layer
> of the matter (p. 625, emphasis added).

With respect to deciding between these two alternatives, the relevance of Wilson's absorption data is that they "*suggest* that the velocity of the rays decreases with thickness of
matter traversed" (pp. 625–626, emphasis added).

While Wilson did not elaborate, the basis of the suggestion is that the absorption
coefficient is not constant but increases in value as the thickness of the absorbing matter
is increased. Since an increase in the absorption coefficient is a measure of the decrease
of the penetrating power of β-rays in making their way through matter, it follows that
those particles suffer a loss of velocity. But this result follows only if the loss of such
penetrating "energy" (as measured by ionization) is not due to other causes whereby
velocity is retained but the number of particles with that velocity decreased after moving
through the media.[26] Thus, the absorption data is only suggestive and not definitive.

Consequently, Wilson decided to conduct "further experiments on this point" (p. 626).
What he did was to *replicate* Schmidt's, 1907 experiment by placing the absorption
screens directly under the electroscope (as in Schmidt's experiment) and then incorporat-
ing his experimental data regarding absorption where the absorption screens were placed
directly behind the radioactive source.[27] Wilson's results are displayed in Fig. 2.11 where:

> Curves shown at *b* and *c* for the rays after passing through 0.489 and 1.219 mm. of aluminium
> were obtained experimentally by varying the field while sheets of aluminium of a given thick-
> ness were placed *under the electroscope*. The rays from the radium were then allowed to
> pass through screens of these thicknesses placed at J *before entering the magnetic field* [as
> in Schmidt's experiment] (p. 627, emphasis in original).

---

[25] Wilson would later go on to demonstrate this experimentally at Gray and Wilson (1910).

[26] Wilson briefly references such theories at (p. 625).

[27] This explains the lack of specific data points for curves *b* and *c* since those curves incorporate
Wilson's earlier results showing linear absorption.

There were two significant results that were obtained in Wilson's replication. First, Wilson's data showed that the maximums for when the β-rays passed freely and then were passed through the absorption screens (when placed at the source), all occurred at the same field strength *just as they had in Schmidt's experiment*. See curves *a, d* and *e* where the maximums all fall on the same field strength. Second, Wilson's absorption data (for when the absorption screen was placed under the electroscope) were displaced to higher field values. See curves *b* and *c*, which show nearly identical maximum values for ionization.

Given these results, it could be determined whether the velocity of β-rays was reduced or remained the same after passing through the absorption screens.

> If the particles do not decrease in velocity in passing through the matter, the curves obtained in this case connecting ionization with strength of magnetic field should [all] fall on *b* and *c*. If the velocity decreases, however, they [i.e., *d* and *e*] should fall to the left of these (p. 626).

Wilson's test turns on the fact that if β-ray velocity does not decrease then it doesn't matter where the absorption screens are placed because (ignoring any differential disturbance on the path to the electroscope) the velocities before and after absorption will be the same, and the reduction in the number of β-rays (after they pass through the absorption screens) will be the same no matter where the screens are placed. On the other hand,

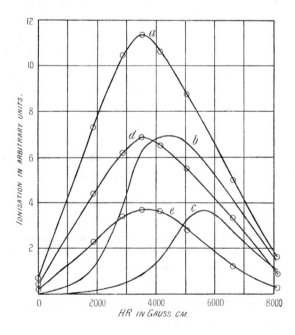

Fig. 2.11 Wilson's velocity-distribution curves, obtained in 1909. *Source* Wilson (1909)

if the velocities decrease then the velocity on curve b (as the β-ray makes its way to the absorption screen under the electroscope) will be greater than that of the β-ray that first passed through the absorption screen before making its way to the electroscope. The experimental result was that curves d and b, and e and c, did not overlap which indicated that velocity decreased. This was on the face of it a very convincing demonstration. Did it warrant *acceptance*, or was it instead only very much *pursuit worthy*?

What though about the coincidence of the maximums for curves a, d and e, where this coincidence was the basis for Schmidt's confident proposal that the β-ray velocities do not decrease as they pass through absorbing media. Thus, there was *apparently conflicting evidence* among Wilson's experimental results. Could this to be *explained*? Wilson thought so:

> This experiment also *explains why* the experiments of Schmidt apparently show no change in the velocity of the rays. According to the views expressed in this paper he was dealing with heterogeneous rays and the position of the maximum should therefore move to the higher fields if the velocity of the rays does not change. The actual decrease in velocity, however, brings the maximum point back to practically the same position as before (p. 627, emphasis added).[28]

As will be seen Wilson on a later occasion elaborated more fully what he had in mind here. And so, we'll postpone our appraisal until we get to the that later elaboration.

Wilson concluded his report with an enumeration of four possible confounding factors: (1) the finite size of the apertures; (2) interference by γ-rays; (3) that, as proposed by J. J. Thomson, "the mechanism of the absorption is not the same for the slow rays... and rapid ones;" and (4) that "[i]t is very probable that if we start with quite homogeneous rays, after they have passed through matter they will become heterogeneous owing to the velocity of each separate particle not being altered by the same amount" (pp. 627–628).

With respect to the confounding effect of the finite size of the apertures Wilson noted that:

> The size of the hole from which the β-rays emerge, and by which they enter the electroscope, makes the beam of rays used in these experiments not quite homogeneous. This would cause the more rapidly moving rays to become relatively more and more important as the rays traverse the matter, and a departure from the straight line should therefore be observed. Experiments made without the [aperture] screens MMM give curves which depart earlier from the straight line. The rays in this case are more heterogeneous, and this is what we should expect (p. 627).[29]

---

[28] What Wilson is referring to here is, as we noted earlier, that Schmidt had expressly noted that his data did not support a pure exponential law of absorption for RaE and were not otherwise explainable in terms of a sum of exponential functions.

[29] See Crowther (1910, pp. 443–444) for a further (but only qualitative) review of the various ways in which non-homogeneous rays could enter through an aperture of finite size.

This is a good argument showing the importance of aperture size since it demonstrates that increasing aperture size makes things worse. But it doesn't show that Wilson's apertures are good enough for his purposes. And as it happened, Hahn and Meitner made the confounding effect of a finite aperture size the centerpiece of their criticism of Wilson's 1909 paper, and it is to those criticisms that we now turn.

2.6   The Problem of the Finite Aperture

Hahn and Meitner did not take Wilson's criticisms sitting down and soon thereafter published their response (Hahn & Meitner, 1909b). After briefly reviewing Schmidt's results they pointedly reminded Wilson what the basis had been for their "working hypothesis" that exponential absorption was a reliable test for the chemical purity of a radioactive substance.

> We have also conducted an investigation of the β-rays of all of the radioactive elements over the last two years and based on our findings we have come to the conclusion that uniform (*einheitliche*) β-rays are absorbed according to an exponential law. *On the other hand, we have never found an exponential law for non-homogeneous β-radiation substances* (p. 948, emphasis added).[30]

Having so established the bona fides of their working hypothesis, Hahn and Meitner cut to the quick:

> Wilson *thinks* that his apparatus provides rays that are sufficiently homogeneous, which he then examines for their absorption (p. 948, emphasis added).

Hahn and Meitner thought otherwise. They began by pointing out that given the size of Wilson's apertures and the radius difference between inner (at 4 cm) and outer radius (at 5 cm), it follows that "the velocities of the beams diffracted into the electroscope by the magnetic field could vary at least in the ratio of 4:5." But this interval is "certainly greater" than that of the velocity of the β-rays emerging from the radioactive source. Which in turn means that "Wilson's calculations, which are based on velocity determinations based on the strength of the magnetic field used and a radius of 4 cm are invalid" (p. 949).

The reader may recall that Wilson claimed that the fact that curves b and d, and curves e and c, do not coincide refutes the Hahn-Meitner hypothesis that β-ray velocity remains unchanged after passing through absorption screens. But according to Hahn and Meitner, Wilson's claim is invalidated because the β-rays in his experiment were heterogeneous due to the finite size of the apertures. Given this heterogeneity, Hahn and Meitner went on to explain why the curves in question did not coincide. To begin "[i]t goes without saying

---

[30] All translations of Hahn and Meitner's response to Wilson are by the authors.

that the maximum for these curves moves to the side of the stronger fields … because the slower rays are for the most part absorbed in the aluminum screen" (949).[31]

At this point we should remind the reader that the points of maximum ionization mark off the *optimized energy value* of particle velocity and number of particles with that velocity. So, as you adjust the field strength to higher values from the left of the maximum you gain energy because of the combination of higher velocity and either a greater number of particles or a lesser number of particles but where the greater velocity outweighs the loss in the number of particles. On the other hand, if you adjust the field strength in a way that moves off the maximum to higher field values you gain higher velocities but fewer particles with the result of a lower net energy than that at the maximum. But this account needs to be augmented by the fact that as you adjust the field strength to mover to higher values of velocity, this causes, as previously noted by Wilson, "the more rapidly moving rays to become relatively more and more important as the rays traverse the matter."

Taking all of this into account, there will be difference in the *relative energy distribution* (between slow and fast rays) of the β-rays that make their way to the electroscope after first suffering absorption and those that do not confront the absorption screen until they get to the electroscope. And given this difference, it follows according to Hahn and Meitner, that "the maximum values for curves *b* and *c* move to the side of the stronger fields" (p. 949).

Hahn and Meitner went on to argue that "Wilson's experiments not only provide no proof of a velocity change, but, when correctly interpreted, lead to the conclusion that the velocity remains unchanged" (p. 949). Hahn and Meitner thus turned the tables on Wilson by arguing that Wilson's claims are invalidated precisely because his β-rays *are themselves heterogeneous*—the very fault that Wilson had argued invalidated the claims of Schmidt, Hahn and Meitner.

Here they focused their attention on the experimental fact (both of Schmidt's experiment *and* that of Wilson) that the maximums for the curves *a, d* and *e all occur at the same field strength*. Hahn and Meitner argued that this cannot be the case given Wilson's theory of a reduction in velocity. Here's why.

In the case of curves *d* and *e*, the absorption screen is placed just below the electroscope. Now assume that J_1 and J_2 represent the intensities (*Intensitaten*) respectively of the slower and the faster rays and that the maximum ionization values for these rays occurs at a field strength where the rays are more fully characterized as $\alpha_1 J_1$ and $\alpha_2 J_2$ where the behavior of faster and slower rays is determined by the ratio α_1/α_1 (p. 950).

---

[31] It is surprising to see Hahn and Meitner appeal to a difference between slower and faster β-rays given their "working hypothesis" that β-rays emitted from pure substances are all of the same velocity and suffer no decrease in velocity when passing through absorbing media. What they may have had in mind is some combination of a relative lack of parallelism (and thus a difference in velocity as opposed to speed) because of the finite aperture size, coupled with other secondary effects such as scatter.

For the Hahn-Meitner theory of unchanging velocity with a variable number of particles remaining after passing through the absorbing media, the J_i's represent the unchanging velocities and the α_i's represent the number of particles involved. The virtue of the notational scheme is that it can also be used to represent Wilson's theory since on that theory the J_i's represent the unchanging numbers of particles and the α_i's the changing velocities.[32] And where, given this notational system, the total absorption of a particle occurs when its velocity is reduced to zero.

Hahn and Meitner next consider what happens when the rays make their way through the apparatus and then pass on through the aluminum screens. For *both* Hahn and Meitner, and Wilson, the absorption can be represented as $\alpha_1 J_1$ changes to $\beta_1 J_1$, and $\alpha_2 J_2$ changes to $\beta_2 J_2$ where $\beta_1 < \alpha_1$, $\beta_2 < \alpha_2$ and $\beta_1 < \beta_2$. Now consider the case where the velocity does not experience any change:

> Then the maximum ionization will always occur at the same field strength as in the case of curve *a*, as long as the ratios β_1/β_2 for curves *d* and *e* are equal to the ratio α_1/α_2 for curve *a* (p. 950).

What Hahn and Meitner had in mind here is that a point of maximum ionization marks an optimized combination of particle velocity and the number of particles with that velocity. Thus, field values near the point of optimization (given the experimentally determined symmetrical shape of the optimization curve) will be similarly symmetrical with respect to *net* energy value.

Thus, their argument is that there's a *possible* combination of parameter values such that the maximums all line up at the same field strength. And that the experimental fact that the maximums do line up indicates that the parameter values in question have occurred. Stated another way, Wilson cannot argue that it is impossible for Hahn and Meitner to account for this alignment of maximums.[33]

Hahn and Meitner next consider Wilson's claim that "the speed is reduced" as the β-rays make their way through the absorption screens.

> In this case there must be a shift of the maximum in the direction of the decreasing field strength *as compared with curve a*, because of the presumed reduction in the velocity after they have made their way through the aluminum screen, and the shift is stronger, the thicker the absorbing screen (p. 950, emphasis added).

---

[32] Actually, there's more than just a notational convenience involved here, since the Hahn-Meitner scheme is also a *simplified account* of a more complex underlying phenomena. Just how much more complex will be seen when we consider (Wilson, 1910).

[33] Hahn and Meitner add the caveat that "[o]nly when you have used so much aluminum that this condition is no longer fulfilled because of excessive absorption, does a shift of the maximum occur, towards the side of the stronger field" (p. 950).

Expanding on this we note that on Wilson's account the velocities of both the fast and the slow rays will be reduced (though the slow rays by a proportionally greater amount since $\beta_1 < \beta_2$. Which means that the mean velocity will be decreased, from which it follows that there will be a shift to a lower field strength which is at variance with the experimental results of both Schmidt and Wilson.

In sum, according to Hahn and Meitner, their account of an unchanging velocity is consistent with the coincidence of the maximums of curves a, d and e, while Wilson's account of changing velocities is not.

How was Wilson to respond? The reader may recall that in his 1909 paper he offered an explanation of how it was that Schmidt could have experimentally determined that the maximums of absorption curves coincided *despite the fact that his rays were heterogeneous and not homogeneous.* And since as forcefully argued by Hahn and Meitner, Wilson's rays were also heterogeneous (because of the finite size of his apertures), this meant that Wilson had an *already made an argument* that could be used to show that his account was also consistent with the experimentally determined maximums. How ironic is that! We earlier postponed a detailed analysis of that argument, to which we'll now turn. As restated in his response to Hahn and Meitner, the argument goes as follows.

> Apparently, Hahn and Meitner have not considered the different strengths of absorption for the slow and the fast β-rays. The postponement of the maximum depends on two factors, firstly from the change in speed of the rays, and secondly from the unequal absorption, which is greater for the slow rays is than for the faster. The first factor causes a shift of the maximum to the side of the lesser field strength, the second one on the side of the larger field strength. Therefore, when the absorbent sheets are located behind the source [at the entrance], we get the joint effect of the two factors (Wilson, 1910, p. 102).[34]

Expanding on this, we note that the "first factor" is just the decrease in velocity of both the slow and the fast rays which means that the *mean* velocity of the rays after passing through the absorption screen has been decreased. If no other factors were involved, a reduced field strength (as compared with that used to locate the maximum for the *a* curve) would have to be used in order to extract the maximum ionization value from this mix of heterogeneous rays.

Wilson argues, however, that there was another factor involved, namely, that the slow rays are extinguished at a higher rate than the fast rays. Which means that a higher field value (than would have been used to extract maximum ionization from the mix of slower rays) would have to be used to achieve maximum ionization. But like Hahn and Meitner who made a similar claim, Wilson does not further elaborate. Moreover, the same justification that we offered for the similar claim by Hahn and Meitner applies, namely, that the deficit created by the elimination of the slower rays can be made up for by moving to a higher field strength—keeping in mind, however, that since maximum ionization depends on an optimization between velocity and number of rays with that velocity there

---

[34] Wilson wrote his response in German and all translations that response is by the authors.

is a limit to how much the field strength could be increased before ionization values decrease.[35]

The complexities of the interaction between Hahn and Meitner, and Wilson provide a telling picture of the problems involved in dealing with the confounding effects due to the use of a finite aperture. As we shall see, these complexities and the associated uncertainties served to motivate Wilson to devise an experiment that would provide for greater homogeneity of the β-rays extracted from the radioactive source and thus a more certain demonstration that the velocities of β-rays are in fact reduced as they made their way through absorbing media. In short, these complexities made the development of a better experimental apparatus a *pursuit worthy venture*. And here we note that a large part of what made the better experiment pursuit worthy is the fact that there was undeniably an element of opportunistic convenience in the accounts offered by Wilson, and as well by Hahn and Meitner. Eliminating the necessity of having to resort to such theory saving moves provided a substantial reason for Wilson to devise more direct and thus decisive experimental proof that the β-particle velocities were reduced after moving through absorbing media.

And as Wilson informed Hahn and Meitner, he had indeed recently completed new experiments where the β-rays produced "are much *more homogeneous*," and which showed that linear absorption is "confirmed with even *greater accuracy* than under the earlier experimental conditions" (p. 102, emphasis added). The question then is how was he able to do this?

2.7   Wilson's 1910 Experimental Response

In order to produce a better experiment, Wilson did not resort to adding more barriers, narrower slits, and longer confined distances to be traveled by the β-rays. Instead, he ingeniously realized that he could achieve better collimation by initially subjecting the β-rays to a magnetic field and then applying a second varying field in order to separate out the β-rays according to their velocity. And so, he accordingly designed and constructed the apparatus shown in Fig. 2.12.

The initial collimation was brought about by the combination of the intervening apertures (at A, L and O) between the source and an aperture at O (where the absorbing substance would be located), and a magnetic field at C that "was kept constant, so that the *same bundle* of approximately homogeneous rays passed through the hole O *during the whole of an experiment*." Once this bundle of β-rays managed to make its way through O it was then subjected to a second magnetic field at D that "was varied" and thereby separated out the β-rays according to their velocity which then made their way to the electroscope at E (Wilson, 1910, pp. 142–144).

---

[35] Wilson also argued (at p. 103) against Hahn and Meitner's claim that "[i]t goes without saying that the maximum for [the curves *b* and *c*] moves to the side of the stronger fields." But for current purposes we need not delve further into those complexities.

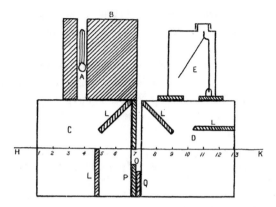

Fig. 2.12 Wilson's refined apparatus of 1910 that yielded data of "greater accuracy than under earlier experimental conditions." *Source* Wilson (1910)

The correlations between field strength and ionization were recorded and presented as curve A as shown in Fig. 2.13. Aluminum screens of different thickness were placed at O, and data collection proceeded by varying the field at D. These results were recorded as curves B, C and D. The soundness of the experimental design was indicated by the fact that "[t]he curves obtained in this manner have well defined maxima" (144).

Fig. 2.13 The data curves obtained with Wilson's improved 1910 apparatus

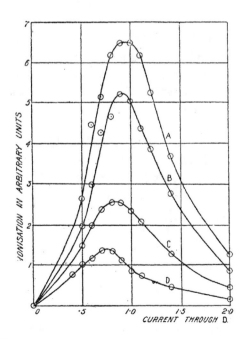

So, the basic idea is that better data would result because the acquisition of that data was made under circumstances that reduced the influence of the finite size of the apertures used. But how much better and compared to what? In order to answer such questions and given the importance of this experiment as a response to the criticisms of Hahn and Meitner, we'll explain the operation of Wilson's experiment in more detail.

To begin, Wilson offered the following observation as a way of providing some insight on finitely sized apertures and the operation of his experiment.

> If the source of the rays and the apertures between the fields and under the electroscope were infinitely small the rays entering the second field would be quite homogeneous, and we would only get ionization in the electroscope for *one definite value of the field in D*. Since the conditions of the experiment require that these should have some finite size, we get instead a curve which rapidly rises to a maximum, which may be taken as a measure of the *mean* velocity of the rays (p. 145, emphasis added).

A similar point could be made if a beam of perfectly parallel rays of the same velocity were projected through the system. Here again, varying D would come up empty except for "one definite value of D."

Add to these abstract considerations the experimental fact that when Wilson varied the field at D, he came up with curve A which indicated that the "bundle of rays" that passed through the system was clearly heterogeneous and to a considerable degree. Ionization varied by a ratio of 7 to 1 (in "arbitrary units"), with a corresponding current variation from 0 to 2.0 A. Nevertheless, Wilson claimed that his apparatus had delivered a "bundle of approximately homogeneous" rays to O. The sense of the expression "approximately homogeneous" in this context is, we suggest, best understood as meaning *less heterogeneous* than that obtained in his 1909 experiment.[36]

To give some substance to this comparison consider the following thought experiment—which must have occurred to Wilson. Begin by separating the right side of the apparatus (from just to the right of O to the electroscope), and then moving the electroscope to just behind O. We now have in essence Wilson's 1909 apparatus (as well as that of Schmidt's, 1907 experiment). Now instead of keeping C at a fixed value, imagine that it's made to *vary* where we record ionization values for the various values of C. In this way we would get a curve of correlations that would be analogous to the *a* curve obtained by Wilson in 1909. Now disconnect the electroscope and reconnect the right-hand side of Wilson's apparatus and move the electroscope back to its original position. With the original configuration thus restored, now carry on with the procedure described by Wilson where we keep the value of C *fixed*, which yields the A curve.

---

[36] In this regard remember that Wilson had earlier informed Hahn and Meitner, that he had recently completed new experiments where the β-rays produced "are much more homogeneous," and which showed that linear absorption is "confirmed with even greater accuracy *than under the earlier experimental conditions*" (Wilson, 1910, p. 102, emphasis added).

Given our thought experiment we now have two curves that record the variation of ionization with field strength: (1) a curve of correlations that is analogous to the *a* curve obtained by Wilson in 1909; and (2) Curve A which yields an analysis (based on the input due to the varying field at D) of the β-particle constituents that correspond to the ionization achieved at a *fixed* field strength at C. Thus, curve A provides an analysis of *the ionization data point (corresponding to value C) on the analogue to* Wilson's 1909 *experiment, where (the area under) Curve A represents the total energy produced at that ionization data point.*[37]

But, as we have seen, curve A reveals the presence of a continuous but heterogeneous, bell shaped distribution of velocities. So instead of thinking that ionization values reflect the presence of rays of *the same or nearly the same velocity*, those values indicate the presence of a *wide continuum* of β-rays that contributed to the experimentally determined ionization value. And where the ionization value is a measure of the mean velocity.

How does any of this help Wilson in his quest to minimize the undue influence of the finite apertures? It helps because the ionization curve arrived at by varying D and keeping C fixed is more focused and bell shaped than the curve that would be arrived at by varying C as in Wilson's 1909 experiment. In other words, *the dispersion of velocities from the mean* is significantly less in Wilson's 1910 experiment. And it is in this sense that Wilson's 1910 experiment delivers a *more homogeneous* set of β-rays to the electroscope than would have been delivered by Wilson's 1909 experiment *using the same radioactive source.* As noted by Rutherford in his review of the experiment:

> Under these conditions an effect in the electroscope is observed over a *considerable range* of current in the second electro-magnet, but the curve has a *sharp maximum* which corresponds to the rays of mean velocity (Rutherford, 1913, p. 242, emphasis added).

We do have to point out, however, that Wilson did not give a formal proof of this comparative improvement. Moreover, any direct comparison of the *a* curve from the 1909 experiment and the A curve from the 1910 was compromised by the fact that different "arbitrary units" were involved. Still, a simple visual comparison indicates that the curve of the 1910 curve A is more sharply defined than that of the 1909 curve *a*. Wilson probably thought as well that the superiority of the A curve over his earlier *a* curve was certainly to be expected given the additional collimation employed and the absence of any evident counteracting, confounding causes induced by the additional collimation.

The experimental and interpretational significance of the superiority of the A curve is that the influence of the "slower" rays is thereby substantially reduced such that the experimentally determined behavior of the β-particles involved is closer to what it would have been if the bundle of those β-particles had been strictly homogeneous. The question

---

[37] For the claim that this area represents the total energy produced at the data point of the curve analogous to curve *a*, see Wilson (1909, p. 622): "The area of this curve, then, represents the ionization we would get in the electroscope if all of the rays of all velocities were allowed... were allowed to enter together instead of being deflected into the electroscope separately by the magnetic field."

then for Wilson was whether his less heterogeneous, aka approximately homogeneous, bundle of rays would yield an experimental manifestation of any significance.

And here the payoff was considerable because *the maximums were no longer in alignment* as they had been in Schmidt's experiment. See Fig. 2.13.

> It will be noticed that these maximum points move to the lower fields as sheets of aluminium are interposed in the path, *proving conclusively* that the velocity of the rays decreases by an appreciable amount as they pass through matter (p. 145, emphasis added).

Wilson's argument in support of his "proving conclusively" claim is just this: That the noncoincidence of the maximums in the direction of weaker field strength cannot be explained if that velocity is assumed not to change.

> The effects observed cannot be due to heterogeneity of the rays which pass into the second field, for if this were the case, and there were no change in the velocity of the rays on passing through matter, the maximum point would move to the higher instead of the lower fields, owing to the fact that the slow rays are more easily absorbed than the rapid ones (p. 145).

If the reader feels a vague sense of *deja vu* that's because Wilson has in effect coopted the argument given by Hahn and Meitner to show that the curves *b* and *c* from Wilson's 1909 experiment shifted to higher field because of "fact that the slow rays are more easily absorbed than the rapid ones." But since the maximums in Wilson's improved experiment move to lower fields this "conclusively" proves that velocity gets reduced when passing through absorbing media.

There were, not surprisingly, a great many experiments conducted in order to replicate and expand on Wilson's results. So, for example, J. A. Crowther constructed an apparatus in which he could measure the velocity of the β-rays before and after they passed through an aluminum absorber. With respect to the question of whether the velocity of β-rays decreased after passing through absorbing media, his results indicated that:

> [T]he absorbing sheet produces a very definite displacement of the curve in the direction of smaller velocities …. It is evident therefore that there is a small, but perceptible decrease in the velocity of the β-rays as they pass through absorbing media (Crowther, 1910, p. 448).

Crowther also investigated the absorption of β-rays concluding, contrary to Wilson, that "the absorption of the rays emitted follows an exponential law." With the proviso, however, that "the absorption of a homogeneous beam of β-rays by a substance such as aluminum which does not emit any large amount of true secondary radiation of its own, follows a law the precise nature of which remains to be determined, but which is certainly not exponential" (p. 456). On the other hand, J. A. Gray on the basis of his experiments concluded that "β-rays, which are absorbed according to an exponential law, are not homogeneous," and that "β-rays must fall in velocity in traversing matter" (Gray, 1910, p. 141).

In his *Radioactive Substances and Their Radiations*, Rutherford tabulated the principal experimental results regarding absorption (Rutherford, 1913, p. 225). Surprisingly, Wilson's 1909 results were not included. Nor were those of Crowther and Gray. Schmidt's results, however, were recognized where his analysis making use of multiple coefficients was noted in three of the entries. Other experiments noted included those by Hahn and Meitner, and Godlewski. In addition, Schmidt's, 1906 results were the first to be discussed in Rutherford's introductory review of absorption experiments (pp. 223–224).

By 1921, as noted by Chadwick in his *Radioactivity and Radioactive Substances*, the consensus was that absorption experiments all fell into one or the other of the following discrete categories.

For many of the products the absorption curve is approximately exponential. For others it is necessary to assume that the curve consists of two or three parts, each of which is exponential with a different coefficient (Chadwick, 1921, p. 52).

And his example of the latter sort of case is in fact Schmidt's, 1906 analysis of radium B. Similarly, Kohlrausch (1928, pp. 190–191) lists 20 examples of experimental determinations of the absorption coefficient where Schmidt's multiple coefficient analysis for RaB and RaC is listed with no mention of Wilson.

Wilson's 1909 and 1910 experiments did, however, get special billing in Rutherford (1913, pp. 234–238, 240–244), and as well in Kohlrausch (1928, pp. 1302–1304). What explains this exclusion from the absorption lists but deserving of a separate focused review? By way of explanation, we note the following distinction made by Rutherford between an *exponential law* of absorption and the *interpretation* of that law.

It must not, however, be forgotten that the *interpretation* of the amount of ionization due to the transmitted rays depends on three factors, (1) the scattering of the β particles back to the side of incidence, (2) the number of β particles stopped by the matter, and (3) the change with velocity of ionization per unit path of the particle. The importance of these factors requires to be settled before any very definite deductions can be made. The apparent absorption according to an *exponential law* of β rays from some radioactive substances is not easy *to account for* on [current experimental] results (Rutherford, 1913, p. 237, emphasis added).

So, for Rutherford, at least, Wilson deserved separate treatment because of the importance of his work for the *interpretation* of the absorption experiments. Rutherford echoed this distinction between *exponential law* and *interpretation* in his later *Radiations from Radioactive Substances* where after briefly describing Wilson's "general explanation" of absorption, and some of the experimental results, he concluded that:

The form of the absorption curve appears to therefore *to have small importance* [for interpretation], since being mainly controlled by scattering it is very dependent on the design of the apparatus and the varying extent to which scattered rays are measured (Rutherford et al., 1930, pp. 414–415, emphasis added).

Thus, as shown by this case, pursuit worthy avenues of experimental investigation will lose their status when their informative value becomes overwhelmed by the confounding effects that have made their presence evident precisely because of that experimental pursuit.

Returning to Wilson's role in all this, we note that by 1912 his research program had taken a different turn where the problem was to determine the "mechanism" of absorption that would explain why rays initially homogeneous became heterogeneous as they made their way through absorbing media. The puzzling experimental result that prompted Wilson's search for such a "mechanism" was, as discovered by Crowther, that:

> [H]omogeneous rays are absorbed by platinum according to an exponential law, and, further, that rays initially homogeneous which have passed through a small thickness of platinum are absorbed according to this law by aluminium (Wilson, 1912, p. 310).

Wilson's proposed explanation turned on a result that he had earlier demonstrated in his 1909 paper, namely, that:

> [I]t is possible to obtain a heterogeneous beam of particles, of which the different types of rays are absorbed according to a linear law, but the absorption of the whole beam takes place according to an exact exponential law (Wilson, 1909, p. 624).[38]

The problem then for Wilson was how to get from that result to his proposed mechanism which was that:

> [I]n their passage through matter some of the particles undergo more violent encounters with the atoms they traverse than others, and there is therefore a tendency for the beam to become heterogeneous. This proceeds until a definite state of equilibrium is reached, when the further passage of the beam through matter produces no alteration in its properties, and the beam is then absorbed according to an exponential law. ... [I]n such a beam the distribution of numbers of the rays composing it with respect to velocity would not alter as it penetrated matter, the greater absorption of the rays of lower velocity being compensated by the decrease in velocity of the more rapid rays, and we should expect experiments to show little or no decrease in average velocity of the rays comprising the whole beam (Wilson, 1912, p. 311).

Wilson's argument in support, as the reader can well imagine, is quite complex but since our immediate interest lies elsewhere, we'll pass over it (though readers are urged to test their mettle and take a look).

We've made this brief review of Wilson's 1912 paper to highlight the fact that the investigation of absorption had evolved into an investigation of how *electrons* interact

---

[38] As restated in Wilson (1912, p. 310): "[T]he exponential law of absorption found for a beam of rays emitted by a single radioactive substance can be explained by assuming that the numbers of particles composing the beam are distributed according to some special law with respect to their velocity."

with absorbing matter rather than with how *β-rays* could be examined in a way that would be revealing as to how they came to be in the first place. One pursuit worthy approach that was taken up in order to reverse this trend was to focus attention on the *beta spectrum* itself and determine more completely and precisely its composition. And so we'll now turn to those experimental efforts.

2.8   A Fork in the Road of Pursuit Worthiness

As we have seen, there were considerable experimental efforts directed at the problem of determining the absorption coefficients for β-particles as they made their way through absorbing substances. But what of the beta spectrum, as reflected in the ionization curve *unadulterated* by interfering absorbing substances? In particular, what was the relationship between ionization and the number of β-particles at an ionization value? The best answer that was available was Paschen's, 1904 experiment.[39] Wilson recognizing the obvious pursuit worthiness of the question, decided to make an experimental determination of this relationship. Knowing this would allow for a more comprehensive specification of the composition of the beta spectrum, as well as its modification after absorption. As summed up by Wilson:

> Although many researches have been made on the properties of the β-rays, no results have so far been published giving the variation of the ionization due to them with their velocity. ...
> A determination of this variation is of importance, since most experiments with the β-rays, e.g., those dealing with their absorption through matter, have been made by measuring the ionization in a vessel through which the β-particles were passing, and such experiments do not give definite information about the actual number of particles involved unless the variation of the ionization of the β-particles with velocity is known (Wilson, 1911, p. 240).

The apparatus employed is shown in Fig. 2.14 where the radium source was located at A, and where a copper cylinder was located at E. A magnetic field was applied with varying intensities in order to extract β-rays whose velocities could be calculated as explained earlier. The experiment was conducted in two stages. In the first, the copper cylinder was insulated and the "negative charge gained by E due to the absorption of the β-particles" was determined as a function of the magnetic field strength and thus the velocities of the β-particles (p. 242).

It is important to note at this stage that what was being measured was the relative change in the *electrical charge* of the brass box and not the ionization produced by the β-particles. The relationship between this charge and the number of β-particles involved is as follows:

---

[39] See also Rutherford (1913, 245–248) for other relevant experimental results.

This charge is proportional to the number of particles entering the vessel, since each particle has the same electrical charge (p. 242).[40]

To determine the relationship between ionization and velocity, the copper cylinder was "replaced by another of exactly the same size and shape, but which was connected to earth instead of being insulated" (p. 243). The experiment then proceeded as before except that it was now the relationship between ionization and particle velocity that was being determined.

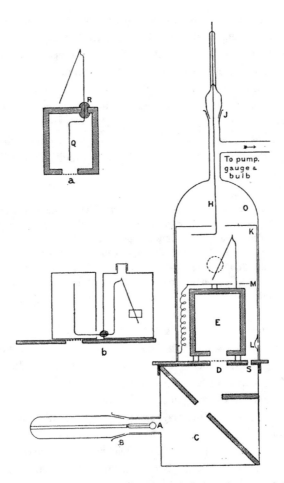

Fig. 2.14 Wilson's apparatus used to determine the correlations between field strength, particle velocity, the relative numbers of β-particles, and ionization. *Source* Wilson (1911)

---

[40] Wilson added: "The actual number of particles was not determined owing to the difficulty of measuring the capacity of the gold-leaf system" (p. 242).

Table 2.2 The correlations between the field strength, particle velocity, the relative numbers of particles, and ionization and also gives the ratio of ionization to charge

Hρ (Gauss cm)	v (10^{10} cm/s)	No. of particles (charge)	Ionisation	Ionisation corrected for reflexion	Ionisation/charge = I	Iv	Iv^2	Iv^3
850	1.35	49	7.67	7.67	15.7	21.2	28	38
1150	1.66	73	9.82	8.31	11.4	20.6	31	52
1390	1.87	87	10.99	8.10	9.3	17.8	32	61
1650	2.08	105	10.97	7.20	6.85	14.3	30	62
1890	2.23	106	10.47	6.50	6.14	13.7	30	71
2160	2.35	108	9.47	5.89	5.45	12.8	30	71
2440	2.46	102	8.64	5.06	4.95	12.2	30	74
2750	2.55	104	7.82	4.66	4.48	11.4	29	74
3300	2.67	89	6.44	3.63	4.08	10.9	29	78
3900	2.74	76	5.43	3.06	4.03	11.0	30	83
4800	2.82	56	3.82	2.16	3.86	10.9	31	87
5430	2.86	47	2.97	1.71	3.64	10.4	30	85
5910	2.88	44	2.58	1.52	3.46	9.9	29	83
6350	2.90	39	2.23	1.37	3.52	10.2	30	86

Source Wilson (1911, 245)

Wilson's experimental results are shown in Table 2.2 which displays the correlations between the field strength, particle velocity, the relative numbers of particles, and ionization and also gives the ratio of ionization to charge. In Fig. 2.15, Wilson graphed the variation of ionization assuming respectively that it varied inversely as velocity, the velocity squared, and the velocity cubed. As can be readily seen, the squared variation was the clear winner.

Thus, Wilson concluded that, "[t]he ionization produced per centimeter by β-particles in free air varies inversely as the square of the velocity between the limits examined" (p. 248). From the table one can also read off the relationship between the ionization produced and the relative number of particles involved.[41]

About this time there was a *reemergence* of a competing experimental apparatus and corresponding methodology for the study of the nature of the beta spectrum. Using the apparatus shown earlier in Fig. 2.1,[42] Von Baeyer and Hahn obtained for ThA + ThD a remarkable photograph where,

---

[41] See Wilson (1912, p. 323) where this relationship was used as part of his explanation of why homogeneous particles became heterogeneous as they made their way through absorbing media.

[42] A rudimentary diagram of this apparatus made its first appearance as Fig. 1 in a separate insert at Von Baeyer and Hahn (1910, p. 520).

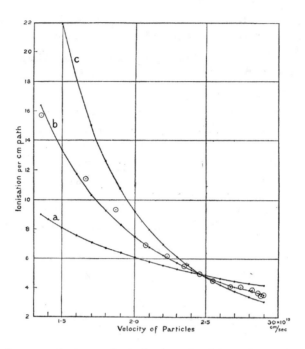

Fig. 2.15 Wilson's measured values and calculated graphs of the variation of ionization assuming respectively that it varied inversely as velocity (**a**), velocity squared (**b**), and velocity cubed (**c**). *Source* Wilson (1911)

the sharp stripe labeled α originates from the α rays, [and where] the more deflected stripes due to ThA are almost as sharp as the original stripe of the α rays, while the weaker ThD stripes, which on the photographic plate are very clearly visible, nevertheless appear in the published reproduction to be barely perceptible. It seems to be beyond doubt that two uniform β radiation groups are present, whose speed is roughly estimated to be about 80 or 95 percent of the speed of light (Von Baeyer & Hahn, 1910, p. 489).[43]

In addition to the lines for ThA and ThD, von Baeyer and Hahn's photographs also indicated the presence of two faint lines about which Baeyer and Hahn stated that they wanted "to leave open what the connection might be between these lines and those of ThA and ThD" (489). By contrast, for radium E they obtained a single, diffuse line, and for mesothorium-2 what appeared to be two partially overlapping lines.[44]

---

[43] Translation by the authors. It's an odd fact that in this early photograph ThA is more strongly represented than ThD, whereas in later photographs, discussed below, the opposite is the case. What explains this apparent inconsistency is the fact that the rays from ThD are not as homogeneous as those from ThA and that this difference was somehow exaggerated in the early photographs.

[44] For an explanation of the identifying terminology for radioactive substances see Rutherford (1913, pp. 534–552).

Unfortunately for reasons unknown, Von Baeyer and Hahn's photograph for ThA and ThD was rather poorly reproduced in the *Physikalische Zeitschrift*. Its "reproduction," as well reproductions of two other photographs, appeared in a separate "Tafel XII" that occurs over 25 pages later (after several other reports), and were all quite small, each measuring only around one-half by three-eighths of an inch.[45] Still despite what must have been their frustration with the reproduction of their photographs, Baeyer and Hahn drew the following conclusion from their results:

> The present investigation proves that during the decay of radioactive substances, not only alpha rays, but also beta rays, leave the radioactive atom with a velocity that is characteristic of the substance in question. Thus, new support has been gained for the assumption made by Hahn and Meitner on the basis of their absorption measurements that such beta rays, which are absorbed according to an exponential law, are homogeneous. Probably, each beta-radiating substance emits only one group of characteristic beta rays. In cases where this does not apply, we may be dealing with complex substances (von Baeyer & Hahn, 1910, pp. 492–493).[46]

The absence of adequately reproduced photographic evidence was rectified in (von Baeyer et al., 1911a). By this time their apparatus had undergone several refinements that improved photographic reproduction including most notably a change of orientation of the photographic plate from perpendicular to the beam path to "approximately parallel to the beam path" such that "one now obtains an image of the entire course of the beam of rays from the slit onwards" (274).[47]

The first of several photographs, shown at Fig. 2.16, highlights the benefits of this rearrangement of photographic plate. Starting from the top right-hand corner and moving counterclockwise, the photograph shows the α particle line, the β particle line for ThD, and two β particle lines for ThA. The photograph also revealed something quite significant, and not all that supportive of the Hahn-Meitner hypothesis of the homogeneity of β-rays. This because as readily can be seen:

> One must draw the conclusion that the [faster] β rays of ThD are not as completely homogeneous as those of [the slower] ThA, even when using the infinitely thin active precipitate as a radiation source (p. 274).

They went on to report the results of a series of absorption experiments on ThA and ThD. In short, after passage through absorbing media, the slower rays showed a decrease in velocity, but preserved homogeneity, while faster rays had only a small change in velocity but became even more inhomogeneous. See, for example, Fig. 2.17 where "in a

---

[45] Because of their small size only a vertical white line is visible against an entirely black background where the vertical line is most likely the α-line and the ThD β line in close proximity. This is the case with the available digitized versions available from the Columbia, Princeton and Michigan University libraries.

[46] Translation by Jensen (2000, p. 27).

[47] Translations from this paper are by the authors.

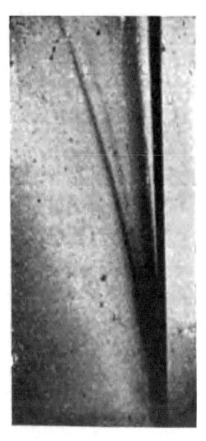

Fig. 2.16 Starting from the top right hand corner and moving counterclockwise, photograph show-ing the α particle line, the β particle line for ThD, and two β particle lines for ThA. *Source* von Baeyer et al., (1911a, 1911b)

simultaneous recording" the upper half of the figure shows the rays after having passed though absorbing media while in the lower half no absorbing medium was interposed.[48]

But there were even more unpleasant surprises in store when von Baeyer et al. soon thereafter found that for RaB + RaC there were five lines from RaB and four from RaC (von Baeyer et al., 1911b). As Hahn later wrote: "Our earlier opinions were beyond salvage. It was impossible to assume a separate substance for each beta line" (Hahn, 1966, p. 57). But even more complexity lay in store.

In 1913 Jean Danysz, using an apparatus that became the basis for a later, exten-sive replication by Rutherford and Harold Robinson, reported the discovery of 27 lines from RaB + RaC (Danysz, 1913). This large increase in the number of component lines

---

[48] Note that the photographic plate had to have been oriented so as to be horizontal as it had been in the earlier experiments.

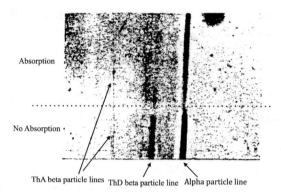

Fig. 2.17 A "simultaneous" photographic recording where the upper half shows the rays after having passed though absorbing media while in the lower half no absorbing medium was interposed. *Source* von Baeyer et al., (1911a, 1911b), with annotations by the authors

attracted Rutherford's attention because uranium X and radium E had so far resisted analysis in terms of a discrete beta spectrum and thereby remained continuous so far as could be determined. And so, Rutherford proposed that:

> In the light of [Danysz's] results it appears not improbable that the continuous β ray spectrum observed for uranium X and radium E may be ultimately resolved into a number of lines. It is quite possible also that the somewhat diffuse band given by thorium D may be caused by the overlapping of several groups of rays (Rutherford, 1913, p. 256).

Thus, replication of Danysz's experiment was pursuit worthy, and not only for that reason, but also because:

> Before any theory of the origin of β and α rays can be adequately tested, it is necessary to determine with the greatest possible precision the velocity of each of the component groups of β rays (Rutherford & Robinson, 1913, p. 719).

Their apparatus, as shown in Fig. 2.18, was a refinement of that used by Danysz and took advantage of the following focusing effect that could be achieved even when using (in order to increase intensity of the β-ray signal) a relatively wide slit or aperture.

> It can be shown theoretically that even with a comparatively wide slit the circles intersect the photographic plate along a curved line of comparatively narrow width. … It is seen from [Fig. 2.18] that if the pencil of rays comprised β rays of only one velocity, the whole pencil of rays would be concentrated along a narrow line on the photographic plate, and consequently the photographic impression for a given exposure would be far more confined and intense than if the photographic plate were placed vertically above the slit where the cone of rays is comparatively wide. It is clear from the figure that this concentration holds for β rays of different speeds (p. 720, emphasis added).

And here we take the opportunity to point out that Fig. 2.19 shows what happens when a particular value of the magnetic field is employed, namely, that β-particles over a relatively wide range of velocities are collected and registered on the expanded photographic plate. This range is indicated by the difference between the rays depicted of larger radius and those of lesser radius, where all intermediate values of velocity would also be included. Moreover, as emphasized by Rutherford, the circular focusing effect applies to all the "β rays of different speeds."

Thus, there's a crucial difference between how the apparatus used for example by Wilson works and how that used by Rutherford and Robinson works. In the first case, a well collimated bundle of rays that will be approximately homogeneous is collected and the ionization values of the β-rays so collected are determined for specific values of the magnetic field. This yields the characteristic bell-shaped data curves. In the second case (and similarly for the incarnation of Becquerel's apparatus used by von Baeyer, Hahn and Meitner) there is no such collimation and the β-rays of a relatively wide range of curvature are collected *all at once* over an expanded photographic plate, where the velocity range of the β-particles registered on the plate depends on the experimental particulars. In short, the distinction is between collecting a large set of individual ionization values for a narrow range of velocities as opposed to collecting all at once the photographic registration over a relatively expanded range of values for velocity. In the next section we'll consider a contradictory, collision between these two methods. But for now, we'll return to the Rutherford and Robinson experiment.

The impressed magnetic field was incrementally varied over a wide range of values so as to include "weak lines" that were difficult to detect. As summarized by Rutherford and Robinson:

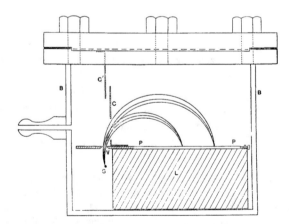

Fig. 2.18 Rutherford and Robinson's apparatus used for photographic registration of the beta spectrum. *Source* Rutherford and Robinson (1913)

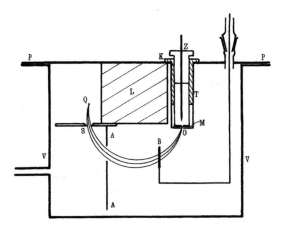

Figure. 2.19 Chadwick's apparatus that led to results that he thought while "very interesting" meant "that the scattering problem will be very difficult." *Source* Chadwick (1914)

In order to examine the whole extent of the β-ray spectrum, photographs were taken in ten different magnetic fields, increasing with approximately constant ratio from about 200 gauss to 6300 gauss. In all, more than fifty separate photographs were taken (pp. 721–722).

Rutherford and Robinson's replication was quite successful, or so it must have seemed, since it resulted in the identification of 64 groups from RaB + RaC where 16 were ascribed to RaB, and 48 to RaC. While Rutherford had contemplated using his refined experimental apparatus to separate the recalcitrant Uranium X and Radium E into their component groups of β rays, a sequence of serendipitous events would have the opposite effect and threatened to relegate Rutherford's hard work to the realm of the instrumental artifact.

2.9   A Challenge to the Reliability of the Photographic Evidence

Around this time James Chadwick as the beneficiary of an 1851 Exhibition Scholarship was off to Berlin to work with Hans Geiger and his newly improved point counter, and in particular to deal with the problem of electron *scattering* which was at the top of the list of bothersome confounding effects. As Chadwick explained in a later 1969 interview:

The one thing I did try to do was to observe the scattering of beta particles, making use of what was then supposed to be a fact, that the beta particles from the active deposit of radium consisted of homogeneous groups of different velocities, quite strong and well separated homogeneous groups (Chadwick, 1969, Session I).

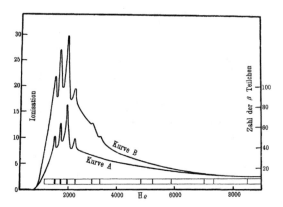

Fig. 2.20 Chadwick's beta-ray spectrum data where Curve A was obtained using Geiger's point counter, and Curve B using ionization determinations. *Source* Chadwick (1914)

But before the thin metal foil could be "bombarded" with β-rays and the scattering effect of the interceding foils thereby ascertained, Chadwick had to determine the number of beta rays in the different groups *before* they ran into the scattering foil. And as a form of initial calibration, he first established "by taking photographs that the β-rays [he was dealing with] were concentrating in sufficiently sharp lines by the magnetic field" (Chadwick, 1914, p. 386).[49] Once he established "the width of the lines on the photographic plate as approximately 0.5 mm," he then proceeded using the apparatus depicted in Fig. 2.19, to count, using the Geiger counter, the number of β-particles as the magnetic field strength was varied. But despite the careful preparation Chadwick found himself stopped in his tracks. As he recounted in a letter to Rutherford:

> I have not made much progress as regards definite results. We wanted to count the β particles in the various spectrum lines of RaB + C *and then to do the scattering of the strongest swift group*. I get photographs very quickly and easily, but with the counter I can't find even the ghost of a line. There is probably some silly mistake somewhere.[50]

But as Chadwick came to realize it was no "silly mistake" but rather the fact that the sought after lines were comprehensively dwarfed by an underlying more or less continuous count of the Geiger counter. Chadwick's results using the Geiger counter were represented in Curve A of Fig. 2.20.[51]

Since the difference in intensity between the peaks and the underlying continuum was only "a few percent" the ionization method was employed just in case by so doing "more

---

[49] Chadwick wrote his article in German, and all translations here are by the authors.

[50] Chadwick to Rutherford, 14 January 1914, cited in Jensen (2000, p. 42), emphasis added.

[51] One peculiarity for which we have no explanation is the fact that Chadwick did not indicate any particular data points on the graphical representation.

lines might be observed." As indicted by the ionization Curve B, these additional "measurements completely confirmed the result using the Geiger counter," and "only the same four lines were observed" (Chadwick, 1914, p. 388).

But as Chadwick realized there was still the possibility that scattering was at least a significant confounding cause of the observed continuum. To eliminate this possibility he "inserted a small lead screen" at location B such which absorbed the primary β-particles from the source but around which the stray and scattered beta rays could make their way and be counted. They were found not to be significant (p. 387).

There was one last thing that Chadwick needed to attend to and that was to *explain* the apparent contradiction between his results and the earlier photographic results. As Chadwick noted at the beginning of his report:

> Based on the [relative] strength of the blackening of the photographic registration, Rutherford and Robinson divided the individual groups of rays [emitted by RaB + C] into seven classes. But since the photographic reaction to β-rays of different velocity is unknown, this is not a reliable method to draw conclusions about the intensity [*Intensität*] of the individual groups of rays. In addition, because of the influence of γ-rays and scattered β-rays, it is difficult to determine by such photographic methods whether or not a continuous spectrum is superimposed over the line spectrum (p. 383).

In other words, the problem was there had not been an effective prior *calibration* of the resultant photographic contrast with the presumably more accurate contrast as indicated by Geiger counter and ionization determinations. Moreover, as made clear by Chadwick's Geiger counter and ionization determinations, the photographic materials and the corresponding exposure and development created an exaggerated increase in contrast which served to overemphasize the line spectrum and to underrepresent the underlying continuum. See pp. 389–390.

Having disposed of this apparent conflict between photographic and non-photographic methods, Chadwick went on to state his final conclusion:

> β radiation presents a continuous spectrum [*kontunuierliches Spektrum*] which is superimposed by a line spectrum of relatively very low intensity [and] it is only in the field of slow β rays that single lines are observable (p. 391).

Understood more expansively Chadwick's result fundamentally changed what it was that had to be explained. What was striking about Chadwick's experimental discovery was not per se the existence of an underlying continuum since that was always to be expected even on the photographic evidence. But rather its substantial intensity which made it seem unlikely that it could be explained away as the result of annoying secondary effects such as scattering and intermeddling gamma radiation.

The physicist Abraham Pais in a concise review of the "earliest experimental studies of the β-spectrum" raised the question of "why Wilson did not [in 1909] go on to prove that the primary β-spectrum is continuous." His answer was that Wilson "probably did not take that step because he was too involved in absorption questions" (Pais, 1986, p. 153). In addition, Pais notes that Wilson could have simply "done the same experiment [as Chadwick] if he had simply omitted all absorbing foils over the slit!" (p. 159). But as a matter of fact, Wilson had done exactly this when he made eight ionization determinations and on this basis constructed what he identified as curve *a* in Fig. 2.11. What Pais apparently had in mind was that Wilson could easily have been more diligent and made additional ionization determinations that presumably would have, as it had for Chadwick, revealed the continuous spectrum.

We think, however, that there is a better answer than that Wilson was "too involved in absorption questions" to make such additional determinations. As we've already noted, Rutherford in 1904 and later in 1913 noted that with respect to the photographic registrations of Becquerel and others, "the impression on the [photographic] plate takes the form of a large diffuse, but *continuous* band" (Rutherford, 1913, p. 98, emphasis added). Moreover, as indicated by Wilson's construction of curve *a* from his experimentally determined ionization values, he must have felt comfortable with the assumption of a continuous range of available velocity values that would match variations in electromagnetic field strength. In short, taking yet more values to establish curve *a* was not called for and hence *was not pursuit worthy*. Its continuity was already *accepted* and *incorporated* into Wilson's data analysis.

Assuming that we're on the right track here, the question that needs to be asked is not why Wilson did not undertake the determination of additional ionization values, but why Chadwick undertook the arduous task of reestablishing what was in effect Wilson's curve *a*. And here it must be remembered that for Chadwick the role of this curve was to provide a background standard to register counting and ionization differences when thin foils would be later introduced and "bombarded" with β-rays, all with the purpose of coming to terms with scattering as a confounding secondary effect. In addition, Chadwick needed to establish the reliability of the Geiger counter which meant that he had to establish that his Geiger counter was able to duplicate the discontinuities in intensity that had presumably been established by the photographic evidence, including his own. But as we have seen he was not able to do so. Eventually he came to realize that he had stumbled upon an underlying continuity on "which is superimposed by a line spectrum of relatively very low intensity" (Chadwick, 1914, p. 391). Which in turn demanded a more thorough and confirming data collection than that made by Wilson.

Chadwick's discovery of the underlying beta continuum does, however, raise the question of how it was possible for not only Wilson but also for his contemporaries, to have missed the combination of underlying continuum with occasional spikes of increased intensity and to instead have managed to collect ionization data that was represented by all those well-defined and continuous bell shaped curves for ionization when no absorbing screens were in play. The answer apparently is that Wilson's ionization values, for example, were as it turned out determinations primarily of the underlying continuum (that had been later revealed by Chadwick) and where the spikes in ionization were too small, too few and of such "very low intensity" to have been discovered given the relatively small number of ionization determinations made by Wilson and others.[52]

Returning now to Chadwick's stated conclusion that "β radiation presents a continuous spectrum which is superimposed by a line spectrum of relatively very low intensity," there is the question of what does this all mean with regard to the production of β-rays? Is the continuous part simply the summation of the various secondary effects in play, or is it indicative of something fundamental about the production of β-rays by radioactive substances? Such questions were without doubt *pursuit worthy* because of their evident centrality to the development of a deeper understanding of the production of beta rays. But pursuit worthy in what way? And based on what *foundation of acceptance*?

For his part, Chadwick's initial reaction was disappointment with his result noting that "this is of course very interesting, but at the same time disappointing, for it means that the *scattering problem* will be very difficult."[53] So at least initially, Chadwick's concern was about the difficulties his newly discovered beta continuum would cause for any explanation of its existence and size based on electron scatter as a confounding cause. In other words, that the continuum was just background noise though at a surprising and alarmingly high level.

But given the relatively large intensity of the continuum (with respect to that of the discontinuous spikes) it had to seem unlikely that the continuum was only the result of secondary confounding effects such as scattering. And following along this vein, that the combination of underlying continuum with superimposed spikes indicates that two separate and distinct causal *underlying fundamental* processes were involved: one for the production of the continuum and another for the production of the spikes. Or at the very least that the investigation of such a possibility was *pursuit worthy*. But World War I interceded, and Chadwick was caught out on vacation in Germany, and was interned, along with the budding engineer Charles Drummond Ellis, for four years while the war continued.[54]

2.10 The Emergence of the Primary and the Secondary Spectrum

Once reunited after the war, Chadwick and Ellis embarked on a replication of Chadwick's, 1914 experiment, and in 1922 they published a report on their results. As they explained,

---

[52] There's another factor that may have come into play here. As noted by Rutherford et al., (1930, p. 399) Chadwick's apparatus "was the first application of the focusing method to electrical measurements" where, as noted above, this focusing method was also taken advantage of in Rutherford and Robinson (1913, p. 720). Consequently, Chadwick's apparatus very likely had greater sensitivity than that used by Wilson. Couple this with the "relatively very low intensity" of the line spikes that Chadwick had been able to discover and it's not surprising that Wilson and other experimenters did not detect those spikes.

[53] Chadwick to Hevesy, 12 March 1914 cited at Jenson (2000, p. 44), emphasis added.

[54] For details of the internment see Chadwick (1969, Sessions I and II).

their experimental apparatus "was identical in principle with that used by Chadwick" in 1914. Radium B and C were again used as the test substances. One noteworthy difference was that a Geiger counter was not used, and they tested only for changes in the ionization as the magnetic field was varied. Other than that, there were no significant differences between the 1922 replication and the 1914 original. Moreover, as they explicitly noted:

> It should be emphasized here that no great accuracy was striven for in these preliminary experiments. Our absolute measurements have an accuracy of about 10 per cent ... (Chadwick & Ellis, 1922, p. 277).

Their results for are shown in Fig. 2.21.

Given the lack of significant difference, the reader may wonder, what was the point of the replication? Keeping in mind the interruption in scientific research and communication caused by the war, the point was in part to reacquaint the scientific community with Chadwick's, 1914 results in a somewhat improved form. More importantly though, the point was to set the stage for their claim that these results provided "strong evidence" for the following:

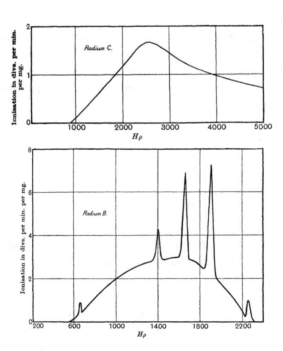

Fig. 2.21 The beta spectrum of radium C (upper) and radium B (lower) when Chadwick and Ellis replicated Chadwick's, 1914 experiment. *Source* Chadwick and Ellis (1922)

[F]irstly, that the continuous spectrum has a real existence which is not dependent on the experimental arrangement and that any explanation of it as due to secondary causes is untenable; and secondly, that our view of the origin of the continuous spectrum is consistent with the magnitude of the observed effects (p. 275).

So unlike Chadwick who in 1914 worried that his experimental results showed "that the scattering problem will be very difficult," the point now was to carefully draw out the consequences of those results for the more fundamental question of how the continuous spectrum was produced.

From their perspective, there were two major contenders for the explanation of the "origins of the continuous spectrum." On the one hand, there was Ellis' proposal that β-particles are born inhomogeneous from an initial disintegration process where:

[These] disintegration electrons form the continuous spectrum, the energy of emission of the electron from any assigned atom being variable within wide limits. The homogeneous groups [i.e., the line spectrum] are considered to be entirely secondary in origin and due to the conversion of γ-rays in the electronic structure of the radioactive atom, these γ-rays being emitted from the nucleus during the disintegration (p. 274).[55]

In other words, for Ellis the continuous part of the spectrum is primary in the sense that it consists of β-particles (of varying velocities) that managed to escape unscathed from the nucleus and through intervening electron orbits to form that continuum. The ionization spikes in the beta spectrum by contrast was made of β-particles that had not escaped unscathed and moreover were ordered into homogeneous groups as to their velocity. Meitner's position was more or less the mirror image. For her:

The β rays expelled from the nuclei of the disintegrating atoms all have the same velocity. One fraction of these β rays passes through the atom unchanged and is measured outside the atom with its full velocity. The other fraction is transformed *in the nucleus* into γ rays of corresponding frequency, and the γ rays eject electrons from the electron rings, which have different velocities depending upon the ionization work performed and which form the secondary part of the β-ray spectrum (Meitner, 1922, p. 132).[56]

Regarding Meitner's account, Chadwick and Ellis went on to claim that:

It would appear that on the [Meitner] theory this continuous spectrum must be considered to be *entirely adventitious* and produced under experimental conditions by some such agency as scattering (Chadwick & Ellis, 1922, p. 274, emphasis added).

It is, however, not immediately clear how this interpretation (of being "entirely adventitious") follows from what Meitner said in the passage from the paper that Chadwick

---

[55] See Jensen (2000, pp. 56–63) for background on Ellis' theory.

[56] Translation by Jensen (2000, p. 64).

and Ellis had cited. The last sentence states that the electrons that go on to make up the continuous part of the spectrum have their origin "from the electron rings" due to a secondary interaction with γ rays. The Chadwick and Ellis conclusion follows only if these secondary interactions occur only "under experimental conditions by some such agency as scattering." But that does not seem to be required by Meitner's position since those interactions could occur earlier in the process of β-particle production. As we shall see, Meitner in a later letter to Ellis clarified what her intentions had been. In any case, what Chadwick and Ellis claimed to be able to show is that:

> [Our] experiments and Chadwick's earlier experiments show *conclusively* that the continuous spectrum is emitted *from the source*, that is, from the brass plate on which is deposited the RaB + C; but they do *not* prove that it arises directly *from the nuclei* of the disintegrating atoms" (p. 278, emphasis added).

Chadwick and Ellis's argument for the truth of the first sentence is that none of the known secondary factors is sufficient to transform electrons with their initially undisturbed energies into a continuum of electrons with different velocities where that continuum matches the size and distribution of the continuous part of the β-spectrum as revealed by Chadwick's, 1914 experiment and the later 1922 replication by Chadwick and Ellis. In more detail, according to Chadwick and Ellis:

> There would appear to be only three ways in which the continuous spectrum could arise in the source. It might be supposed to consist of electrons ejected from the material of the source by the γ-rays; or it might consist of electrons which originally formed part of the homogeneous groups, but which had been rendered heterogeneous by being scattered back from the brass plate; or, lastly, the continuous spectrum might be emitted as such by the radioactive atoms (p. 278).

Chadwick and Ellis then argued that the first possibility was "ruled out at once by the magnitude of the effect," which was too large to be caused by γ rays. The second possibility was eliminated by measuring the same spectrum for a source deposited on a very thin silver substrate. They found that only twenty percent of the effect could possibly be due to scattering from the brass plate. They also found that the ratio of the peaks to the continuous background was the same in both the silver and brass substrate experiments. That would be expected only if the original emitted spectrum was continuous. For details see pp. 278–279.

What was left then was the third, and only remaining possibility, that "the continuous spectrum [is] emitted as such by the radioactive atoms" (p. 278). Thus, as summarized by Chadwick and Ellis:

> In our opinion these experiments strongly support the view that the continuous spectrum is emitted *by the radioactive atoms themselves*, and any theory of the β-ray disintegration must take thus into account (p. 279, emphasis added).

Nevertheless, as noted earlier, Chadwick and Ellis added the qualification that their arguments "do *not* prove" that the continuous spectrum "arises directly *from the nuclei* of the disintegrating atoms" which was the central component of Ellis' theory of β-particle creation. But how was one to get inside the nucleus to determine that Ellis was correct about this? Ellis' response was to appeal to the simplicity of his proposal.

> Under these circumstances our hypothesis that the continuous spectrum consists of the actual disintegration electrons [from within the nucleus] seems to be the simplest way of viewing the facts (p. 279).

But if so, then:

> This hypothesis could be put to [experimental] test … for on this view the number of electrons emitted per second in the continuous spectrum, say of radium *B*, should be equal to the number of atoms of RaB disintegrating per second (p. 279).

Ellis' argument here is best understood as claiming that the fact that his hypothesis "is the simplest" justifies the *pursuit worthiness* of developing an experimental test, where that pursuit worthiness is reinforced by his suggestion of a strategy for developing an effective experimental test. Chadwick and Ellis thus concluded their paper with an extended discussion of how it could experimentally be shown that "the number of electrons emitted per second in the continuous spectrum [is] equal to the number of atoms of RaB disintegrating per second." See pp. 279–280.[57]

Like Chadwick and Ellis, Meitner also thought that a replication of Chadwick's, 1914 experiment was in order, that is, was pursuit worthy. But unlike them, Meitner had in mind a *significantly improved version* of that experiment. And so, under her direction, her student Werner Pohlmeyer went on to do exactly that. The principal innovation was the use of much larger ionization chamber which resulted in correspondingly better resolution as shown by his results. See Fig. 2.22 where in addition to the ionization results a representation of the photographic exposures of ThB and ThC + C" decay was included as the small insert. Thus, Pohlmeyer was able to calibrate and thereby demonstrate the consistency between photographic 2 and ionization methods of determining the intensity of the β-rays as a function of the magnetic field strength.

As can be readily seen there were two striking differences when compared with Chadwick's, 1914, as well as with Chadwick & Ellis', 1922 results.

> First, in our graph not only strong groups, but also weaker ones appear; second, the large maxima dominate over a relatively weak background, whereas for Chadwick the maximum intensities exceed their close vicinity by at most 60 percent (Pohlmeyer, 1924, p. 224).[58]

---

[57] Ellis was not the only one who considered such an investigation pursuit worthy. For an extensive review of the experimental efforts with this purpose in mind see Jensen (2000, pp. 123–128).

[58] Translations by Jensen (2000, pp. 92–93).

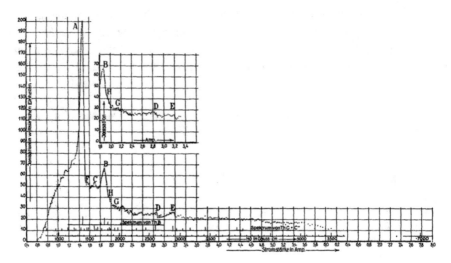

Fig. 2.22   Pohlmeyer's (ionization) beta spectrum for ThB + C, where a graphical representation of the photographic exposures of ThB and ThC + C" is inserted in the figure. *Source* Pohlmeyer (1924)

Based on his results, Pohlmeyer made the case that the features of the beta spectrum were explainable as the *combined result* of three factors: (1) the limited resolution of both his and Chadwick and Ellis's experimental apparatus; (2) scattering of the beta particles at the level of the radioactive wire source; and (3) inhomogeneities caused by collisions within the radioactive atoms (pp. 227–228).

Pohlmeyer concluded by claiming the results of the Chadwick's experiments "can in no way be advanced in favor of Ellis's point of view, according to which the primary β rays from the nucleus caused by the disintegration do not emerge with a uniform velocity" (p. 229). But that's a conclusion that, as we have seen, Chadwick and Ellis also drew when they expressly noted that their experimental results "do not prove that [the continuous spectrum arises] arises directly from the nuclei of the disintegrating atoms" (Chadwick & Ellis, 1922, p. 278).

The dispute between Ellis and Meitner is more fairly and accurately described by noting that they both agreed that the crux of the matter lay within the nucleus and the surrounding electronic orbits, and on that there was not yet any resolution as to whether the β-particles were born with varying velocities already in hand or whether they were all born with the same velocity which then became varied because of interactions within the nucleus and the electron orbits. Moreover, they both agreed that secondary effects within the experimental apparatus played a role in the creation of the appearance of a continuous spectrum but where Ellis and Meitner disagreed on the extent of those secondary effects.[59]

---

[59] See. e.g., Chadwick and Ellis (1922, p. 276) where they argue that the "effect of stray β-radiation, arising from scattering" is "of small importance to the order of accuracy of the experiment."

With regard to her views on such secondary effects, we note that in 1925 Meitner sent the following comment by letter to Ellis.

> I always held the opinion that on leaving the nucleus, the primary β rays must have a well-defined velocity. But already in my second paper [of 1922] I drew attention to the fact that, by excitation of the characteristic radiation in the mother atom as well as because of the arbitrary collision conditions, the β rays can lose any amount of energy, and so must become inhomogeneous. ... It is therefore a mistake [to] say that I maintain the view that the continuous background is due in the main to insufficient resolution of the lines [i.e., secondary instrumental effects] only. That this circumstance must contribute something to the continuous spectrum is clear, but I held this effect to be decisive only in my first paper.[60]

In sum, there was agreement that the investigation and development of accounts of the production of β-particles within the nucleus and the surrounding electronic orbitals constituted in the main the content of what was pursuit worthy with respect to the production of the continuous spectrum.

2.11 Denouement: The Ellis and Wooster Heat Experiment

By 1925 Ellis and Meitner had proposed competing theories as to the initial creation of β-particles within the atoms of radioactive substances. For Meitner, β-particles arose with the same velocity out of an unspecified internal conversion process within the nucleus that preserved energy and quantization. To explain the difference between the discrete line spectrum and the less energetic continuum, Meitner proposed a variety of secondary interactions either within the electron orbits or in the experimental apparatus whereby some of the initially created β-particles escaped unscathed (forming the line spectrum) while others were involved in secondary interactions with γ rays within the confines of the atomic orbits or with other particles while passing thorough the measuring apparatus, thus forming the less energetic continuum part of the spectrum. Thus, for Meitner, β-particles came before everything else.

Ellis, on the other hand, had proposed that beta particles arose already equipped with varying velocities out of a complex set of initial interactions with γ-particles and electrons within the nucleus or the electron orbits with the result that some of those electrons eventually emerged from the nucleus as β-particles with varying energies that corresponded to the line and the continuous parts of the beta spectrum.[61]

But by 1925 a number of experiments had been conducted that had the cumulative effect of decisively showing that the production of β-particles occurred before the production of γ-particles and not the other way round. On this there was agreement

---

[60] Meitner to Ellis, 16 November 1925, cited at Jensen (2000, p. 78).

[61] See Jensen (2000, pp. 55–119) for an extensive account of the development of these and other theories as to the internal workings of the atoms and nuclei of radioactive substances.

and acceptance, including by Ellis and Wooster who after reviewing these experiments concluded that:

> We may sum up by saying that four different and independent experiments each lead to the same result, that the γ-rays are emitted after the disintegration (Ellis & Wooster, p. 852).

While these experiments effectively put an end to Ellis' specific account of the production of β-particles they did not eliminate the possibility that beta particles arose by some other process in an initial state whereby they already had varying velocities. In particular Ellis realized that he could avoid the problem caused by the discovery that γ-rays are emitted after the disintegration that led to the creation of β-particles by taking advantage of the accepted experimental fact that "Radium E… emits no γ-rays and its entire emission consists of a continuous spectrum of electrons without any definite groups" (p. 858). And because Radium E emits no γ-rays the question of beta or gamma ray priority no longer matters. In short, nature itself afforded a ready-made example of a simplified system of β-particle production. But how was Radium E to be experimentally probed in a way that would decide whether in this case beta particles were all born with equal velocities (as held by Meitner) or with varying velocities (as held by Ellis).

Ellis realized that in his search for an experiment that would decide the issue he could make use of the then accepted experimental fact that:

> [T]he number of electrons emitted agrees almost exactly with that to be expected if it were formed by the disintegration electrons, that is, *one electron for each atom disintegrating*. … It seems unnecessary to labour this point and we may accept the fact that *disintegration electrons* form a continuous spectrum (p. 858, emphasis added).[62]

In sum and applying all this to RaE, it follows that each disintegration produces a single electron, i.e. a β-particle, and where those electrons (postulated by Ellis to be inhomogeneous as to velocity) form the continuous spectrum *directly* without the intervention of γ-rays or other forms of interaction. And since Radium E does not produce any discrete (higher energy) line spectrum there is no need to explain anything other than the continuous spectrum. Perfect!

Well almost. Here the problem was that this admittedly "provisional working hypothesis" does not explain "how this inhomogeneity of velocity has been introduced" (pp. 855–856, 858). An immediate explanation would be that conservation of energy fails in the disintegration process. Bluntly stated, that it's a fundamental fact about the world that β-particle production is a process that violates the conservation of energy. But this was rejected from the start.

---

[62] Here Ellis and Wooster made reference to Emeleus (1924, p. 400).

> We assume that energy is conserved exactly in each disintegration, [even though] if we were
> to consider the energy to be conserved only statistically there would no longer be any diffi-
> culty in the continuous spectrum. But an explanation of this type would only be justified when
> everything else had failed, and although it may be kept in mind as an ultimate possibility, we
> think it best to disregard it entirely at present (p. 858).

Therefore, despite the temptation of a simple and immediate "explanation" in terms of
the failure of the conservation of energy, Ellis and Wooster decided instead to stick with
the conservation of energy as a constraint on the problem of devising an experiment
using Radium E that would distinguish between Ellis's newly rejuvenated theory of the
creation of *inhomogeneous* β-particles and Meitner's contrary proposal that all β-particles
are initially created with the same velocity and thus are *homogeneous*. How then were the
two proposals to be tested experimentally? Answer: Determine *the difference between the
energy expended* in the creation of β-particles according to the two competing accounts.

This difference could be determined in the following way. Since the continuous spectrum data for β-decay displays the range of final velocities for the β-particles one can
determine the energy expense given Meitner's account by considering the difference
between the average energy of the β-particles and the energy produced by a β-particle
that had managed to escape without intervening loss. So, for RaE:

> The continuous spectrum of radium E extends to about 1,000,000 volts, showing a maximum
> at about 300,000 volts, and it is a question of accounting for an average loss of at least 500,000
> volts per atom disintegrating if initially every electron is emitted with the maximum velocity
> (p. 859).

Since Meitner's proposal was that all β-particles initially arrive on the scene with identical
velocity, this means that the high-end value of 1,000,000 V can be taken as a measure of
the energy of β-particles that have not incurred secondary interactions that result in a loss
of energy. But as the beta spectrum reveals, not all of Meitner's postulated β-particles
are so fortunate as to have arrived on the scene unscathed by secondary interactions of
some sort of other. Given the range of final velocities, that average loss in energy loss is
around 500,000 V per β-particle. Assuming that Meitner is correct, the question then is
what accounts for this *specific* average deficit. For Ellis, there is no such question since
his β-particles are born with the varying velocities indicated in the beta spectrum. That,
of course, as Ellis realized, does not explain how his β-particles came to have such a
range of velocities.

Staying focused on Meitner's account, Ellis proceeded to consider how her β-particles
had their velocities varied *once they had exited the nucleus* where these proposals included
collisions as the they made their way through the rings of orbiting electrons, or further
down the line when on their way to the electroscope. But all these proposals were rejected
on the grounds that their likely effects would be too small to account for the existence of
the energy loss as indicated by the β-particle spectrum. See p. 859.

Now since, according to Ellis and Wooster, there were no acceptable explanatory mechanisms that would account for the 500,000-V energy deficit it follows that:

> We are thus left with the conclusion that the disintegration electron is actually emitted *from the nucleus* with a varying velocity. We are not able to advance any hypothesis to account for this but we think it important to examine what this fact implies (p. 859, emphasis added).

In other words, since there are no acceptable explanations *in terms of secondary interactions outside of the nucleus*, it must be the case that β-particles are "emitted *from the nucleus* with a varying velocity." But if so, what are the consequences of such a conclusion?

One such hypothesis to consider, what Ellis and Wooster refer to as "the most obvious assumption" is that "even if the energy with which the disintegration electron is ejected varies from atom to atom, yet *the total energy given out in each disintegration is the same*" (p. 860, emphasis added). Such a hypothesis relies on the fact that Ellis and Wooster had only rejected possibilities for energy depleting interactions *with the escaping β-particles*. What was still left was some sort of *simultaneous production* of β-particle and energy in *the initial disintegration process* that occurred *within the nucleus*. In other words, there is the "obvious assumption" that the production of β-particles is *monoenergetic* in the sense that the sum of the kinetic energy of a β-particle and its postulated accompanying energy (in some form or other) is exactly equal to the kinetic energy of a β-particle that makes its way into this world with the maximum velocity. What are the advantages and disadvantages of such a proposal?

> This would make it possible to consider the nucleus in a completely quantized state both before and after the disintegration, but it brings with it the necessity for finding the missing energy. The amount of this missing energy is large, in radium E it would be more than 500,000 volts per atom, and as we have seen already it is very difficult to find any quantum mechanism to provide it (p. 860).

So, while monoenergetic creation retains consistency with fundamental quantum mechanical requirements for quantization, it creates the experimental problem of determining the "missing energy" that accompanies the creation of β-particles during the disintegration process as it occurred within the nucleus. If it could be shown that there was no such missing energy then Ellis' proposed theory that β-particles are, as it were, born directly with varying velocities would have survived such an experimental test.

The good news in all this, as Ellis explains, is that there is no need to "enter into these difficulties" of searching for such missing energy "because there is a *direct* way of finding out whether the above assumption [that β-particle production is monoenergetic] is true or not." And that is, "to find the heating effect of the β-rays from radium E" (p. 859, emphasis added). We will describe this experiment in detail shortly, but for now it's enough to know that the basic idea is to encase the RaE in an insulated enclosure and let the radioactive decay generate an amount of heat that will be equal the total energy produced by the β-particle disintegration process. Thus:

> If the energy of every disintegration is the same [i.e., monoenergetic] then the heating effect should be between 0.8 and 1.0×10^6 volts per atom and the problem of the continuous spectrum becomes the problem of finding the missing energy. It is at least equally likely that the heating effect will be nearer 0.3×10^6 volts per atom, that is, will be just the *mean kinetic energy* of the disintegration electrons (p. 860, emphasis added).

The unattractive consequence, however, of this second possibility is that:

> This would make it impossible for the nucleus to be rigidly quantized both before and after the disintegration and would denote a lack of definiteness of which there has hitherto been no evidence (p. 860).

This is because, assuming conservation of energy, and working backwards from the varying (and thus non-quantized) velocities of the β-rays (as postulated by Ellis), it follows that the initial disintegration events will also be non-quantized. So, either way, on Ellis and Wooster's analysis, the heating experiment will lead to a highly problematic result: Missing energy that's nowhere to be found or giving up basic principles of quantum mechanics. And here we note that consistent with their earlier statement that "[w]e assume that energy is conserved exactly in each disintegration," Ellis and Wooster have not raised the possibility that conservation of energy has been violated, only that quantization was at risk.

Before proceeding to a discussion of the heating experiment itself, we note that Ellis and Wooster concluded with the following argument made on behalf of conducting the proposed experiment.

> We have discussed these possibilities solely with the object of emphasizing the interest and importance of the continuous spectrum. There seems to be no doubt that it exists and that the explanation of its occurrence is not to be sought in any ordinary secondary effect. Some very interesting phenomenon seems to be involved in the β-ray disintegration, and the first step towards its elucidation is the determination of the heating effect of radium E (p. 860).

In other words, that a heating effect experiment is *pursuit worthy* for the reasons given, which include the *promise* of a result that will likely be revealing as to the explanation of the existence of the continuous beta spectrum. But here's an implicit corollary: Being pursuit worthy does not mean that if there is acceptance at the end of the road, that there will not be a new and more menacing set of theoretical and experimental challenges.

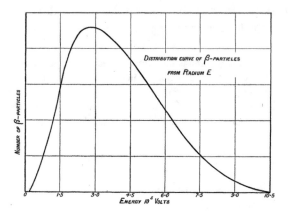

Fig. 2.23 Madgwick's updated determination of the RaE spectrum. *Source* Ellis and Wooster (1927)

It was not until 1927 that Ellis and Wooster were finally able to complete and report the results of their proposed heating experiment.[63] By then a more accurate determination of the RaE spectrum had become available. See Fig. 2.23.[64]

On this updated basis the average energy of disintegration turns out to be "about 390,000 V" (Ellis & Wooster, 1927, p. 111). Where, as the reader may recall, this is a measure of the energy required to produce β-particles if they are assumed to be produced with a varying kinetic energy as postulated by Ellis and Wooster. On the other hand, if one assumes a monoenergetic creation of β-particles, then since β-particles "are emitted with energies as high as 1,000,000 V" the total energy per monoenergetic creation of a β-particle (i.e., the sum of the kinetic and what Ellis and Wooster now refer to as the "characteristic" energy) "cannot be less than this figure, and atoms which emit the slower electrons must get rid of their surplus energy in some other form" (pp. 111–112).

In addition to making use of an updated determination of the RaE spectrum, Ellis and Wooster presented a somewhat simplified account of the theoretical options where the choices were now presented as being between:

(1) "the disintegration electrons are actually emitted from the nucleus with a definite characteristic energy as in the case of the α-particles," or
(2) "disintegration electrons must be emitted from the nucleus with varying energies" (pp. 109–110).

---

[63] For a brief review and analysis of such heating experiments see Rutherford (1913, pp. 568–586). We note as well that since Ellis and Wooster (1925a) had earlier conducted a heat experiment, they already had some experience in conducting such experiments.

[64] From Ellis and Wooster (1927, p. 111) citing "experiments of Mr. Madgwick carried out at the Cavendish Laboratory" (p. 120).

As further explained and summarized by Ellis and Wooster:

> On a previous occasion (Ellis & Wooster, 1925b) we have discussed the secondary effects that might reasonably be expected to occur, and we showed that were these effects to be present with sufficient intensity to account for the inhomogeneity, then simple experiments would already have given direct evidence of their occurrence. It was on these grounds that we concluded that the disintegration electrons must be emitted from the nucleus with varying energies, however contrary at first sight this might appear to be to the general principles of the quantum theory. This conclusion is so fundamental for the whole subject of β-ray disintegration, and has been the occasion of so much controversy, that *it is highly important to have more direct proof* (p. 110, emphasis added).[65]

And here's how the amount of this surplus energy could be more directly determined by a determination of the amount of heat produced by the production of β-rays. While the basic concept is simple—put the radioactive substance in a teapot and see how long it takes the water to boil—the particularities involved made for a very difficult experiment.

> It is well known that no large amount of penetrating radiation is emitted by radium E, so that if this hypothetical surplus energy really does exist, it must be absorbed inside the calorimeter and will contribute to the heating effect. In this case the heating effect would be 2 [to] 6 times as great and would correspond to 1,000,000 volts per atom (pp. 111–112).

Given, as we shall see, the considerable difficulty in actually conducting such an experiment, and the consequential expense in likely experimental error, it was essential for its feasibility and thus for its pursuit worthiness, that there was such a large difference in the predictions made by the competing hypotheses of creation with varying velocity versus monoenergetic creation of velocity and accompanying surplus energy. Thus, as stated by Ellis and Wooster:

> It will be seen that a measurement of the heating effect provides a unique distinction between the two hypotheses, since one predicts a value of 390,000 and the other 1,000,000 volts per atom (p. 112).

Because only small sources of RaE were available, Ellis and Wooster had to prepare a calorimeter that could reliably determine temperature increases that were within 10^{-3} °C.[66] As explained in their 1925 paper, the central purpose of the heat experiment was to determine the heating effect of pure RaE. But since Ellis and Wooster "were never able to prepare a source entirely free from polonium" (p. 112), this meant that

---

[65] Note though that the possibility of monoenergetic disintegration within the nucleus (as discussed in the 1925b paper) is not mentioned. This omission may be because the "missing energy" would be the same for monoenergetic disintegration within the nucleus as it would be for secondary causes acting on β-rays once they had escaped the nucleus.

[66] For a detailed description of the calorimeter see (p. 113).

the calorimeter could only measure the heating effect of the combination of RaE and its polonium decay product.

The experimental challenge was to find a way to extract from the available temperature determinations that part of the heating effect that was due to the RaE. And once this was done, there was still the problem of somehow extracting from the total heating effect the average heating effect per disintegration. But because the decay and subsequent heating remained sealed during the 20 or so days that it took to complete an experimental run, there was no feasible way of counting number of β-particles emitted from the RaE. Here's a step a by step analysis of how Ellis and Wooster were able to overcome these challenges in an experiment that's been deservedly called both "beautiful" and a "tour de force."[67]

Step 1: As will be seen, an essential part of the experimental design required knowing the heating effect of pure RaE *before it began to decay into Po*. But since Ellis and Wooster "were never able to prepare a source entirely free from polonium" they had to calculate how long it would have taken a pure source (if it had been available) to decay into the combination of RaE and Po that was in fact used at the beginning of the experimental measurements. This was accomplished by determining ("by an ordinary α-ray ionization measurement") what the output of α particles of the Po was at the beginning of the experiment as well as at the end of the four experiment runs, which were about 20–25 days later.[68] Since the decay constants (i.e., the half-lives) of RaE and Po were known, it was possible to "calculate the time at which the source *would have been pure radium*." (p. 112, emphasis added) According to their calculations this "was found to be 2 days before the first α-ray measurement was made" (p. 117). See Fig. 2.24 which displays the results from what Ellis and Wooster described as their "best experiment" of the four performed. We have annotated this figure to show where the zero point for time was located, which was around two days to the left of the time of the first reported experimental determination.

Step 2: For the next 20 days or so the RaE + Po "source was left in the brass tubes in the calorimeter and not touched in any way, and, during this time, [seven] heating determinations were made" (p. 115). We've marked these measurements with small circles along the topmost curve labelled "Total Heating."

Step 3: The next step was to extrapolate the topmost curve to what its value would have been at time zero. Here, this value was estimated to be about 25 mm of heat effect (112). This value occurs at the top left corner of Fig. 2.24.

Step 4: Since the half-life of RaE was known to be 5.0 days, the amount of RaE present at the end of the experiment runs (around 26 days) was calculated to be "0.027 of its initial value." If the heat effect varied as the amount of RaE, this meant that the heating effect was equal to 25 mm times 0.027 which is equal to 0.675 mm. If this value is

---

[67] Meitner 1929 letter sent to Ellis cited in Jensen (2000, p. 130), and Pais (1977, p. 927).

[68] Po only produces alpha particles and moreover it was known that such particles were emitted all with the same velocity.

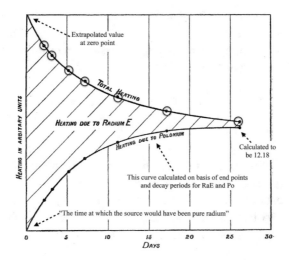

Fig. 2.24   The heating curves obtained by Ellis and Wooster. 1927 *Source* Ellis and Wooster (), with annotations by the authors

then subtracted "from the measured heating of 12.85 mm, we obtain the heating due to the polonium formed" which is 12.18 (117). In Fig. 2.24 this value is indicated at the righthand side of the curve designated "Heating Due to Polonium."

Step 5: Note that we now have the two end points (at time 0 and 26) for the heating due to Po. Using these end points and extrapolating "backwards, using periods of 5.0 days and 139 days respectively for radium E and polonium, we obtain the lower curve, showing the heating due to the polonium" (p. 117).[69]

Step 6: Once the location of the Po heating curve has been established by the above extrapolation, one can simply read off the heating values (as the difference between the Total Heating and the Heating Due to Polonium curves) at the times when the initial *experimental* determinations were made of the total heating effect. Moreover, "[t]he difference between the two curves *must* show the heating effect due to radium E" (p. 117, emphasis added).

What, the reader may wonder, justifies the "must." Answer: Given the procedures for extrapolation and coordinated curve construction, the only available space for occupancy by "the heating effect due to radium E" is that designated as Heating due to Radium. But now the question to ask is: Does *must* mean that this area *in fact* fully captures the

---

[69] Both decay periods were used in order to determine the *net* half-life of the Po which takes into account the contribution due to RaE decay and the radiation loss on the part of the Po. Assuming that the resulting amount of Po is proportional to the heating effect (as had been earlier assumed for the RaE), one then in effect inserts this extrapolation into Fig. 2.24 where the extrapolation shows "the heating effect due to the polonium" (p. 117).

heating effect due to Radium E? In other words, what guarantees that there is no escape of heat from the calorimeter? If there were, then the experiment would be inherently biased against the Meitner hypothesis of monoenergetic production of β-particles. In response to this concern, Ellis and Wooster note that:

> [I]t is a most important confirmation of the accuracy of our experiments that this difference [between the TOTAL HEATING and the HEATING DUE TO POLONIUM curves] shows an exponential decay with a period of about 5.1 days (p. 117).

This counts as a confirmation of accuracy because the experimentally determined result of 5.1 days closely matches the then accepted decay period of 5.0 days for radium E. In short, what counts as evidence in favor "of the accuracy" of the Ellis and Wooster experiment is the fact that the complicated procedure employed by Ellis and Wooster in measurement and analysis yields a value (5.1) that so closely matches that acquired by accepted independent methods.

This completes the process that went into the creation of Fig. 2.24. For all the times when experimental determinations were made (in their "best" experiment), the experimentally determined values as well as the calculated values were collected into the first four columns of Table 2.3.

The question then becomes how does one make a determination of the "Disintegration energy of Radium E" (the last column) on the basis of the values for *total heating* and the proportional contributions due to RaE. And here we note that by the "Disintegration energy of Radium E" Ellis and Wooster mean the average energy expended in the creation of individual β-particles. If one knew the number of disintegrations at each of the times in question, it would be a simple matter of dividing the total heat due to RaE by the

Table 2.3 The average energy of disintegration for RaE, taken at different decay times

True age (days)	Total heating (mm)	Portion due to Po	Portion due to radium E	X	Disintegration energy of radium E In volts
2.25	22.0	3.68	18.3	15.4	339,000
3.20	20.8	4.91	15.9	15.5	337,000
5.20	19.0	6.99	12.0	15.5	337,000
7.20	17.8	8.64	9.2	15.6	335,000
11.20	16.1	10.53	5.6	14.5	360,000
17.20	14.2	11.83	2.4	14.7	355,000
26.20	12.85	12.18	0.67	15.1	346,000

Source Ellis and Wooster (1927, 111)

number of the β-particles emitted from the RaE. But given the design of the calorimeter there was no practical way of counting these particles.

Here Ellis and Wooster made good use of considerable algebraic creativity and a small bit of underlying theory to effectuate the determination of the average energy of disintegration despite not having a direct count of the β-particles emitted by the RaE. Here's how they did it. First note that:

The heating effect of RaE at time t will be equal to: (the number of RaE disintegrations) X (the average energy given out at a RaE disintegration).

Similarly:

The heating effect of Po at time t will be equal to: (the number of Po disintegrations) X (the energy given out at a Po disintegration).

And here we need to point out that the energy given out at a Po disintegration will be a fixed value because Po emits alpha particles of uniform velocity, and where the value of the corresponding energy was known to be 5.22×10^6 V.

Now letting λ_E and λ_P represent the decay constraints for RaE and Po, Ellis and Wooster note that:

If at zero time there are N_E radium E atoms and no polonium, then after time t there will be $N_E \lambda_E e^{-\lambda Et}$ radium E disintegrations, per second and $N_E \lambda_E \lambda_P (e^{-\lambda Pt} - e^{-\lambda Et})/(\lambda_E - \lambda_P)$ polonium disintegrations per second (p. 117).

Finally, for notational convenience they let x be equal to:

(the energy given out at a Po disintegration) divided by (the average energy given out at a RaE disintegration).

Putting this all together, it follows that at time t, *the ratio* of the heating effect of radium compared with the heating effect of Po will be equal to:

$$[e^{-\lambda Et}/(\lambda_E - \lambda_P)]/x\lambda_P(e^{-\lambda Pt} - e^{-\lambda Et})$$

And here we note that the use of ratios solves the problem of not knowing the number of β-particles involved because the $N_E \lambda_E$ coefficients cancel out.

The value for x can now be determined because the values for the λ_i's were known, and all other terms were known on the basis of the experimental results. So, for example, as indicated in Table 2.3, at time 2.25 days, the portion of heating due to Po was 3.68, while that due to RaE was 18.3, which led to a value of 15.4 for x. If 5.22×10^6 V, the energy of disintegration of polonium, is now divided by this value of x, then one gets a value of 339,000 for the disintegration energy of RaE in volts. Similarly, for the other columns in Table 2.3 which indicated the results for seven days ranging from the beginning to the end of the experiment. With these results in hand, Ellis and Wooster drew the following conclusion:

> Considering the nature of the experiment, the agreement is excellent and shows that the energy of disintegration of radium E cannot be much different from 344,000 volts [the average of the Disintegration column] (p. 118).

This somewhat qualitative conclusion, however, was not the end of the story since Ellis and Wooster went on to give an extensive analysis of the likely confounding factors, and on the basis of that analysis made an estimate of the likely error in their determinations of disintegration energy as given in the last column of Table 2.3.[70] On the basis of this analysis, Ellis and Wooster stated their final conclusion as follows:

> The above considerations clearly indicate that the electrical methods show that the average energy per disintegration of the emitted particles is 400,000 volts to within 15 per cent., and this is in good agreement with the average total energy of disintegration found by the heating method of 350,000 volts ± 40,000 volts.
>
> We may safely generalize this result obtained for radium E to all β-ray bodies, and the long controversy about the origin of the continuous spectrum of β-rays appears to be settled (p. 121).

Given its theoretical importance and we expect personal importance to Meitner, it's not surprising that she decided to replicate the Ellis and Wooster experiment. And here we want to take the opportunity to make a simple but fundamental point about such replications. There is by and large no point in exactly repeating an experiment. That would just repeat whatever experimental faults or shortcomings that existed in the first place. Such repetition only makes sense if there's good reason to think that there was a confounding variable that was likely to change in value if the experiment were repeated. It's more efficient to construct a better experiment that is specially designed to address suspected shortcomings and confounding effects. Stated another way, what's typically pursuit worthy is the better experiment and not a strict replication. A straightforward example of such pursuit worthiness is Wilson's 1910 replication of his earlier 1909 experiment where the changes were made to better collimate the bundle of rays that were to be subjected to analysis by means of making variations in the second field at D.

---

[70] See pp. 118–121. While we have chosen, in the interests of economy of presentation, to skip over this material (and related appraisals of confounding factors in the experiments discussed earlier), we want to emphasize that a central feature of every experiment is an analysis of the effect of confounding factors, where this analysis typically involves experimentation on the experimental apparatus itself. Acceptance and even pursuit worthiness depend on the results of such analyses of confounding factors. We'll go over this aspect of experimentation in more detail in our next case study.

So too with respect to Meitner's replication that was designed with several improvements in mind. Most important was the fact that she and her collaborator Wilhelm Orthmann were able to more directly determine the number of decaying RaE atoms per second and thus avoid the complex procedure that Ellis and Wooster had to employ. This was made possible because they had the means and skill needed to prepare a RaE source that was significantly purer than that used by Ellis and Wooster. In addition, Orthmann had developed a calorimeter that was more robust and less sensitive to disturbance than that used by Ellis and Wooster.[71]

What they found was an average β-particle energy of 337,000 electron volts with a claimed accuracy of $\pm 6\%$. The corresponding value for the Ellis and Wooster experiment was 344,000 electron volts with a claimed "overall accuracy of at least 10 percent." So, the two experimental results were very close, and given the very large difference in predicted disintegration values, *close enough* to justify *acceptance* of the general experimental result, as stated by Ellis and Wooster, "that in a β-ray disintegration the nucleus can break up with emission of an amount of energy that varies within wide limits" (Ellis & Wooster, 1927, p. 121). But what were the implications of such acceptance for energy conservation and quantized values at the nuclear and atomic level? Meitner very neatly combined gracious acceptance and a consideration of consequences in a subsequent letter sent to Ellis in 1929.

> It seems to me now that beyond any reasonable doubt you are completely right in your hypothesis that the nuclear β rays are primarily inhomogeneous. However, I cannot at all understand this result. We have very carefully searched for a possible continuous γ radiation, but only a much too weak γ radiation is present. And yet, as long as we are not prepared to abandon the energy law, there is no theory which would not demand a continuous γ radiation equivalent to the continuous β spectrum. According to quantum mechanics, too, such a γ radiation should be there; however, it seems to be present neither for RaE nor for ThC. I am very anxious to learn the solution of this riddle (cited at Jensen, 2000, p. 142).

Ellis and Nevill Francis Mott gave the following concise appraisal of the status of these still unresolved consequences as of 1933.

> The difficulties connected with the continuous β-disintegration are well known. The fact that a given isotope of any element has a definite atomic weight suggests that the energy of the normal state of any nucleus is quantized; further evidence is afforded by the alternating intensities in band spectra. Evidence from the fine structure of α-rays and from the γ-rays, proves that nuclei are capable of existing in quantized excited states. *In fact, in all transformations where α-particles, γ-radiation or protons are ejected from nuclei, the evidence suggests, (i) that the nuclear energy is quantized, and (ii) that energy is conserved.* On the other hand, when a nucleus P transforms itself into a nucleus Q by emission of a β-particle, the β-particle has

---

[71] For more details on these improvements and on the additional efforts by Meitner and her colleagues to detect confounding causes that would result in heat leaking from the calorimeter see Jensen (2000, pp. 141–143).

all energies between zero and a definite upper limit. *One may either conclude that the energy either of P or of Q is not quantized, or that energy is not conserved in the transition.* Since the α-transitions leading up to P, and starting from Q, show no sign of any indefiniteness in the energy, it is difficult to accept the former alternative; and it is thus usual to suppose that energy is not conserved (Ellis & Mott, 1933, p. 502, emphasis added).

For their part in adjudicating whether to abandon conservation of energy or quantization, Ellis and Wooster, in a final section to their 1927 report, proposed an ingenious conglomeration of classical and quantum elements. The basic idea was that while quantization ruled in the outer belts of orbiting electrons, there was an unquantized inner electronic region whereby interaction with the outer belts it could be the case that "the final statistical result [would] follow regular laws, whether the real life of the nucleus is entirely ordered or not" (Ellis & Wooster, 1927, pp. 122–123).

Niels Bohr, on the other hand, wanted to close the door on the intrusion of such classical concepts into the atomic realm, and enthusiastically championed the abandonment of energy conservation (both statistically and absolutely) and the creation of a radically new physics. Wolfgang Pauli more out of desperation than conviction proposed a solution to the "missing energy" problem by proposing the existence of a new particle, which he dubbed "neutron",[72] that was able to surreptitiously carry off the surplus energy to parts unknown.[73] This neutral, light particle was supposed to exist in the nucleus and was emitted along with the electron in beta decay.[74] The existence of such a particle would solve the problem of the "missing energy" and by so doing explain the continuous electron energy spectrum.[75]

---

[72] After the discovery of the neutron by Chadwick in 1932, a heavy particle that was considered to be a constituent of the nucleus, there was some confusion in the physics literature between the two particles, Chadwick's and Pauli's. Fermi solved the problem by christening Pauli's particle the "neutrino," or little neutral one.

[73] For more on Bohr and Pauli, and other responses to the puzzle of the missing energy, see Jensen (2000, pp. 145–184) and Franklin (2001, pp. 51–151).

[74] At the time the accepted model of the atomic nucleus was that it consisted of protons and electrons, the only two massive particles known. In addition to the energy conservation problem in beta decay, this model also had difficulties with the stability of the nucleus, the size of nuclear magnetic moments, and the spins of several nuclei.

[75] Pauli's suggestion also solved a problem associated with the spins of several nuclei. However, other problems remained. These were solved when Chadwick discovered a neutral particle with a mass approximately that of the proton. This neutron was incorporated into the model of the nucleus and used in Fermi's theory of beta decay. For Fermi beta decay was the decay of a neutron into a proton, electron, and a neutrino. In such a three-body decay the electron is no longer required by conservation of energy and momentum to have a unique energy. For details see Franklin and Marino (2020, pp. 60–62).

In sum, there was no consensus on what was uniquely pursuit worthy, only that there was a range of such possibilities. Agreement to be sure that something was amiss, but exactly what was the question. Thus, while there may have been agreement on the general experimental result of the Ellis and Wooster experiment, that agreement brought no resolution as to its consequences. This is where we will end our story.

References

Becquerel, H. (1900a). Contribution a l'etude du rayonnement du radium. *Comptes Rendus des Seances de L'Academie des Sciences, 130*, 206–211.

Becquerel, H. (1900b). Sur la dispersion du rayonnement du radium dans un champ magnetique. *Comptes Rendus des Seances de L'Academie des Sciences, 130*, 372–376.

Becquerel, H. (1900c). Note sur la transmission du rayonnement du radium au travers des corps. *Comptes Rendus des Seances de L'Academie des Sciences, 130*, 979–984.

Becquerel, H. (1901). Sur l'analyse magnetique des rayons du radium et du rayonnement secondaire provoque par ces rayon. *Comptes Rendus des Seances de L'Academie des Sciences, 132*, 1286–1288.

Bragg, W. H. (1904). On the absorption of alpha rays and on the classification of alpha rays from radium. *Philosophical Magazine, 8*, 719–725.

Chadwick, J. (1914). Intensitatsverteilung im magnetischen Spektrum der β-Strahlen von Radium B + C *Verhandlungen der deutschen physikalischen Gesellschaft, 16*, 383–391.

Chadwick, J. (1921). *Radioactivity and radioactive substances*. Sir Isaac Pitman and Sons.

Chadwick, J. (1969). Interview of James Chadwick, 1969 April 15. Niels Bohr Library & Archives, American Institute of Physics. C. Weiner.

Chadwick, J., & Ellis, C. D. (1922). A preliminary investigation of the intensity distribution in the β-ray spectra of radium B and C. *Proceedings of the Cambridge Philosophical Society, 21*, 274–280.

Crowther, J. A. (1910). On the transmission of β-rays. *Proceedings of the Cambridge Philosophical Society, 15*, 442–458.

Danysz, J. (1913). Recherches experiment ales sur les rayons β de la famille du radium. *Annales de Chimie et de Physique, 30*, 241–320.

Ellis, C. D., & Wooster, W. A. (1925a). Note on the heating effect of the γ-rays from RaB and RaC. *Mathematical Proceedings of the Cambridge Philosophical Society, 22*, 585–586.

Ellis, C. D., & Wooster, W. A. (1925b). The β-ray type of disintegration. *Proceedings of the Cambridge Philosophical Society, 22*, 849–860.

Ellis, C. D., & Wooster, W. A. (1927). The average energy of disintegration of radium E. *Proceedings of the Royal Society (London), A117*, 109–123.

Ellis, C. D., & Mott, N. F. (1933). Energy relations in the β-type of radioactive disintegration. *Proceedings of the Royal Society (London), A141*, 502–511.

Emeleus, K. G. (1924). The number of β-particles from radium E. *Proceedings of the Cambridge Philosophical Society, 22*, 400–403.

Franklin, A. (2001). *Are there really neutrinos?* Perseus Books.

Franklin, A., & Marino, A. (2020). *Are there really neutrinos*. CRC Press.

Gray, J. A. (1910). The distribution of velocity in the β-rays from a radioactive substance. *Proceedings of the Royal Society London [A], 84*, 136–141.

Gray, J. A., & Wilson, W. (1910). The heterogeneity of the β rays from a thick layer of radium E. *Philosophical Magazine, 20*, 870–875.

Hahn, O. (1909). Über eine neue Erscheinung bei der Aktivierung mit Aktinium. *Physikalische Zeitschrift, 10*, 81–89.

Hahn, O. (1966). *Otto Hahn: A scientific autobiography*. Charles Scribner's Sons.

Hahn, O., & Meitner, L. (1908a). Uber die Absorption der β-Strahlen einiger Radioelemente. *Physikalische Zeitschrift, 9*, 321–333.

Hahn, O., & Meitner, L. (1908b). Uber die β-Strahlen des Aktiniums. *Physikalische Zeitschrift, 9*, 697–704.

Hahn, O., & Meitner, L. (1909a). Nachweis der komplexen Natur von Radium C. *Physikalische Zeitschrift, 10*, 697–703.

Hahn, O., & Meitner, L. (1909b). Uber das Absrorptionsgesetz der β-Strahlen. *Physikalische Zeitschrift, 10*, 948–950.

Jensen, C. (2000). *Controversy and consensus: nuclear beta decay 1911–1934*. Birkhauser.

Kohlrausch, K. W. F. (1928). Radioaktivitiit. *Handbuch der Experimentalphysik, Band 15*. Akademische Vedagsgesellschaft m.b.H.

Meitner, L. (1964). Looking back. *Bulletin of Atomic Scientists*, 2–7.

Meitner, L. Z. f. P. (1922). Über die Entstehung der β-Strahl-Spektren radioaktiver Substanzen. *Zeitschrift fur Physik, 9*, 131–144.

Pais, A. (1986). *Inward bound*. Oxford University Press.

Paschen, F. (1904). Über die Kathodenstrahlen des Radiums. *Annalen der Physik, 319*, 389–405.

Pohlmeyer, W. (1924). Über das β-Strahlenspektrum von ThB + C. *Zeitschrift Fur Physik, 28*, 216–230.

Rutherford, E. (1899). Uranium radiation and the electrical conduction produced by it. *Philosophical Magazine, 47*, 109–163.

Rutherford, E. (1904). *Radio-activity*. Cambridge University Press.

Rutherford, E. (1913). *Radioactive substances and their radiations*. Cambridge University Press.

Rutherford, E., Chadwick, J., et al. (1930). *Radiations from radioactive substances*. Cambridge University Press.

Rutherford, E., & Robinson, H. (1913). The analysis of the β rays from radium B and radium C. *The London, Edinburgh, and Dublin Philosophical Magazine and Journal of Science, 26*, 717–729.

Schmidt, H. W. (1906). Uber die Absorption der β-Strahlen des Radiums. *Physikalische Zeitschrift, 7*, 764–766.

Schmidt, H. W. (1907). Einige Versuche mit β-Strahlen von Radium E. *Physikalische Zeitschrift, 8*, 361–373.

Von Baeyer, O., & Hahn, O. (1910). Eine neue β-Strahlung bein Thorium X; Analogein in der Uran- und Thoriumreihe. *Physikalische Zeitschrift, 11*, 448–493.

von Baeyer, O., Hahn, O., et al. (1911a). Uber die β-Strahlen des aktiven Niederschlags des Thoriums. *Physikalische Zeitschrift, 12*, 273–279.

von Baeyer, O., Hahn, O., et al. (1911b). Magnetische Spektren der Beta-Strahlen des Radiums. *Physikalische Zeitschrift, 12*, 1099–1101.

Wilson, W. (1909). On the absorption of homogeneous β rays by matter, and on the variation of the absorption of the rays with velocity. *Proceedings of the Royal Society (London), A82*, 612–628.

Wilson, W. (1910). The decrease of velocity of the β-particles on passing through matter. *Proceedings of the Royal Society of London, A84*, 141–150.

Wilson, W. (1911). The variation of ionisation with velocity for the β-particles. *Proceedings of the Royal Society of London, A85*, 240–248.

Wilson, W. (1912). On the absorption and reflection of homogeneous β-particles. *Proceedings of the Royal Society (London), A87*, 310–325.

The Wu Experiment

3

3.1   Parity and Symmetry Transformations

We start with a simple question: What are the spatial transformations that define the symmetry between original and reflection in a simple bathroom mirror? While there's an initial temptation to think that the x coordinate goes to $-x$, it's actually the front-to-back z value that gets sent to its negative value. Nature, at least when it comes to classical physics, is more demanding because all the laws of classical physics—mechanics and electrodynamics—are such that their application outcomes are symmetrical with respect to what's known as the parity transformation, where x, y and z get mapped into their respective negative values, $-x$, $-y$ and $-z$.

With one possible exception this is also true if you also take into account quantum physics. Interactions involving the strong, electromagnetic and gravitational forces are all well behaved in the sense of being equivalent under the parity transformation. This is very handy since it allows for a test for what processes are permissible and which are prohibited: If the parity transformation is not equivalent to the original then the process is prohibited.

In the case of quantum physics all of this gets encapsulated in the concept of the *parity* value of a subatomic entity, a particle or system of particles. In short—and avoiding the rather complicated formal details—the parity of such entities is a measure of the transformational properties of the associated wave function. Here it's convenient to shift to polar coordinates whereby the parity transformation is represented by the transformation of the vector **r** describing a position in three-dimensional space to its directionally contrary $-$**r**. Thus, if $\psi(\mathbf{r})$ is the wave function of a particle, or a system of particles, where **r** is the vector describing its position in three-dimensional space, then, if $\psi(\mathbf{r}) = -\psi(-\mathbf{r})$ the wave function has *odd or negative parity*. If $\psi(\mathbf{r}) = \psi(-\mathbf{r})$ the parity is *even or positive*. Continuing in this vein, the parity of a system of particles, turns out to be equal to $(-1)^l$

where l is the orbital angular momentum or spin. The wave functions of different particles may behave differently under the parity transformation. We refer to this as the intrinsic parity of the particle. The parity of an isolated ensemble of particles is the product of the spatial parity of the system and the intrinsic parities of the particles.

Now we get to the principle that controls (or was thought to control) all allowable physical processes in the quantum realm.

Conservation of parity: The parity of an isolated ensemble of particles after interaction will be equal to the parity of the system before interaction.

Finally, and tying all of this together, there is a fundamental connection between *parity conservation* and *the parity coordinate transformation* which is captured by the following theorem.

If a decay process is not invariant under the parity transformation, then parity conservation does not hold.[1]

Roughly speaking, in order for parity conservation to hold, the parity transformation of a process must be congruent to the physical process as represented in the original system of spatial coordinates. If for convenience we refer to the original coordinate system as the "real space" and that space as modified by the parity transformation as the "parity space," then the process must look the same in both spaces in order for parity to be conserved. When we get to the actual Wu experiment we'll explain how this theorem applies and what being congruent (looking the same) in the relevant respects comes to. The good news in all this is that for present purposes, namely, obtaining a basic understanding of the Wu experiment, it's enough to know this theorem.

We mentioned above that interactions involving the strong, electromagnetic and gravitational forces are all equivalent under the parity transformation. There was some question, however, whether processes involving the weak force were also so well behaved. What brought this question to the fore was the following puzzle involving two K mesons.

3.2   The Theta-Tau Puzzle and a Proposed Solution

Based on their different modes of decay, the theta meson and the tau meson were apparently different particles.

$$\Theta^+ \rightarrow \pi^+ + \pi^0$$
$$\tau^+ \rightarrow \pi^+ + \pi^+ + \pi^-$$

But because they had the same mass, charge and lifetime, the natural conclusion is that *they are the same particle* and happen to have two different decay modes. But there's a problem if we assume conservation of parity. Assuming that the spin of θ and τ is 0, then the parity of the θ decay, $\pi^+ + \pi^0$, is:

---

[1] For the proof of the theorem see Gibson and Pollard (1976, pp. 119–127, 160–162).

$$(-1)(-1) \times (-1)^L = (-1)(-1) \times (-1)^{\text{spin}=0} = (-1)^2 \times (1) = +1$$

where the (-1) terms represent the intrinsic parity of the pi meson, $(-1)^L$ the orbital angular momentum parity of π^+ and π^0, and $(-1)^{\text{spin}=0}$ the orbital angular momenta parity of θ. Conservation of angular momentum requires that the orbital angular momenta of $\pi^+ + \pi^0$ be equal to that of θ, which means that $L = 0$.

Similarly, the following yields the parity of the τ decay, $\pi^+ + \pi^+ + \pi^-$:

$$(-1)^3 \times (-1)^{L_1+L_2} = (-1)^3 \times (-1)^{\text{spin}=0} = (-1)^3 = -1$$

Therefore, given *conservation of parity*, it follows that θ and τ are really *two different particles* despite having the same mass, charge, and lifetime.[2] Hence, the puzzle.[3]

In 1956, Tsung-Dao Lee and Chen-Ning Yang realized that the puzzle could be solved if parity were not conserved in the weak interactions, the interaction responsible for radioactive decay and for the decay of many elementary particles. After examining the existing evidence for parity conservation, they found, to their surprise, that although there was good evidence for parity conservation in the strong and electromagnetic interactions, there was, in fact, no evidence available for the weak interactions. They then offered their radical solution to the problem.

> One way out of the difficulty is to assume that parity is not strictly conserved, so that θ^+ and τ^+ are two different decay modes of the same particle, which necessarily has a single mass value and a single lifetime. We wish to analyze this possibility in the present paper against the background of the existing experimental evidence of parity conservation. It will become clear that existing experiments do indicate parity conservation in strong and electromagnetic interactions to a high degree of accuracy, but that for the weak interactions (i.e., decay interactions for the mesons and hyperons, and various Fermi interactions) parity conservation is so far only an extrapolated hypothesis unsupported by experimental evidence (Lee & Yang, 1956, p. 254).

Lee and Yang went on to consider the evidential status of the puzzle itself.

> One might even say that the present $\theta - \tau$ puzzle may be taken as an indication that parity conservation is violated in weak interactions. This argument is, however, not to be taken seriously because of the paucity of our knowledge concerning the nature of strange particles. It supplies rather an incentive for an examination of the question of parity conservation. To decide unequivocally whether parity is conserved in weak interactions, one, must perform an experiment to determine whether weak interactions differentiate the right from the left (Lee & Yang, 1956, p. 254).

---

[2] For more details and the theoretical support for these parity determinations (including the role of the $L_1 + L_2$ superscript) see Gibson and Pollard (1976, pp. 159–160).

[3] For the current status of the puzzle see Bellantoni (2016).

In terms of acceptance and pursuit, the possibility of a solution to the puzzle was certainly not sufficient for acceptance of the proposition that parity was not conserved, but given the centrality of parity conservation, that possibility was sufficient for experimental pursuit. Or so it was judged to be by Lee and Yang, and ultimately by Chien-Shiung Wu and the other members of what became her experimental team. This despite the fact that such non-conservation would be a definite outlier in a field of parity conservation.

3.3   A Proposed Experiment

Having proposed that "parity conservation is violated in weak interactions," the problem was to devise an experiment that would "determine whether weak interactions differentiate the right from the left" (Lee & Yang, 1956, p. 254). And in fact, Lee and Yang came up with several such proposals. The simplest and most direct was to determine whether β decay from an oriented, i.e., polarized nucleus, was distributed in a way that did not satisfy the parity transformation requirement.

> A relatively simple possibility is to measure the angular distribution of the electrons from β decays of oriented nuclei. If θ is the angle between the orientation of the parent nucleus and the momentum of the electron, an asymmetry of distribution between θ and $180° − \theta$ constitutes an unequivocal proof that parity is not conserved in β decay. If the angular distribution were symmetrical that would argue for parity conservation.... To be more specific, let us consider the allowed β transition of any oriented nucleus, say Co^{60}. The angular distribution of the β radiation is of the form

$$I(\theta)d\theta = (\text{constant})(1 + \alpha \cos \theta)) \sin \theta d\theta$$

where α is proportional to the interference term .... If $\alpha \neq 0$, one would then have a positive proof of parity nonconservation in β decay.... It is noteworthy that in this case the presence of the magnetic field used for orienting the nuclei would *automatically cause a spatial separation* between the electrons emitted with $0 < 90°$ and those with $0 > 90°$. Thus, this experiment may prove to be quite feasible (p. 255, emphasis added).

More simply stated: the idea is to somehow orient a sample of Co^{60} such that the β-ray emissions go more or less straight up or straight down, and to then test for the equality of the numbers of emissions that go up versus those that go down. If there's equality, then parity is conserved. Otherwise, it's not conserved.

Before proceeding we note that there was significant skepticism that any experimental actualization of Lee and Yang's experimental proposals would show nonconservation of parity for weak force interactions. In a letter to Victor Weiskopf, Pauli wrote, "I do not believe that the Lord is a weak left-hander, and I am willing to bet a very large sum that the experiments will give symmetric results" (Bernstein, 1967, p. 59). Similarly, Richard

Feynman bet Norman Ramsey, both Nobel Prize winners, \$50 to \$1 that parity would be conserved. Feynman paid (Feynman, 1985, p. 248). Felix Bloch, yet another Nobel Prize winner, offered to bet other members of the Stanford Physics Department his hat that parity was conserved. He later remarked that it was fortunate he didn't own a hat (Lee, 1985).

3.4   Actualization of the Proposed Experiment

We now turn to the question of how the "relatively simple possibility" proposed by Lee and Yang was to be actualized in a real-world experiment. Enter Chien-Shiung Wu, Ernest Ambler, Raymond Hayward, Dale Hoppes, and Ralph Hudson. First on their agenda was how to orient, i.e., polarize, the Co^{60} nuclei in a way that could be both *monitored* and *isolated* from systemic interference. As we shall see, a determinate and stable polarization state was necessary in order to show that the parity transformation was not congruent with the real-world state. In this regard it was known that "Co^{60} nuclei can be polarized by the Rose-Gorter method in cerium magnesium (cobalt) nitrate, and the degree of polarization detected by measuring the anisotropy of the succeeding gamma rays" (Wu et al., 1957, p. 1413). In short, the method of polarization was to prepare certain structurally advantageous salts containing Co^{60} that under significant cooling could be polarized by the imposition of a magnetic field. We'll explain the importance of the "succeeding gamma rays" in the next section.

In addition to the general problem of achieving and maintaining the very low temperatures required, there was the problem that "the radioactive nuclei must be located in a thin surface layer and polarized" (p. 1413). As can be seen in Fig. 3.1, the thin crystalline layer containing Co^{60} on the upper surface of a single crystal of CMN was located at the lower end of the apparatus, known as a "cryostat", while a thin anthracene crystal, located above the Co^{60}, was used to detect the beta particles where the scintillations were transmitted to a photomultiplier.

Initially, the experiment was attempted without the housing with the result that the specimen very quickly heated and "[t]he polarization lasted no more than a few seconds, then completely disappeared" where:

> The reason for this disappearance of nuclear polarization on the surface was probably due to its sudden rise in temperature caused by heat that reached the surface of the specimen by means of radiation conduction or condensation of the He-exchange gas. The only remedy was to shield the thin CMN crystal in a cooled CMN housing (Wu, 2008, p. 55).[4]

---

[4] This reference contains a reprint of a 1983 lecture by Wu.

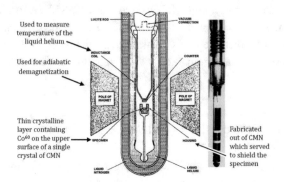

Fig. 3.1 The experimental apparatus of Wu et al. *Source* Hudson (2016), augmented by authors

So, a housing had to be designed that would insulate the specimen long enough to allow for the collection of reliable data. That turned out to be a very arduous and time-consuming task especially since the housing had to be assembled with individual crystal components.[5] But once production of the component parts was completed, they were glued together using DuPont ("Duco") cement, which at the time was a nearly universally used bonding agent. And so, with the housing completed and installed along with the specimen, the magnets were turned on and then disaster: the housing "came tumbling down!" As Wu later recounted:

> In making the housing, one must line up the crystal axis perpendicular to the demagnetization field and glue the CMN pieces together. [But because the] axis of the crystal had not been set exactly parallel to the magnetic field, a strong torque developed, the ultra-low temperature caused the DuPont cement to completely lose its adhesive property; and the CMN housing under the torque came tumbling down! (pp. 56–57).

It was at this stage that someone on the team thought to make use of "fine nylon threads to tie the pieces together" (Wu, 2008, p. 57). See Fig. 3.2.[6] And so the power was again turned on. This time with great success. Not high tech, but a simple yet effective solution. The collected data on the beta particle counts were presented in the form as shown in Fig. 3.3 and revealed, for many, the shocking result that:

> [T]he emission of beta particles is more favored in the direction opposite to that of the nuclear spin (Wu et al., 1957, p. 1414).

---

[5] See Wu (2008, pp. 55–56) for just how difficult this turned out to be.

[6] David Christen (Oak Ridge National Laboratory) informs us that the current product of choice used to secure wayward components when conducting experiments near absolute zero is waxed dental floss. It behaves exactly as you'd want.

Fig. 3.2 Ernest Ambler's notebook for the experiment, Dec. 27, 1956: "tie crystals in bundle" and "parity not conserved!" *Source* Hudson (2016)

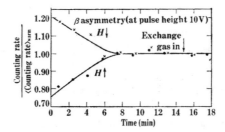

Fig. 3.3 Collected data for "beta particle asymmetry". *Source* Wu et al. (1957)

And just to be clear, we note that because the apparatus could only measure the upward moving β particles, the experimental determination of the discrepancy between upward and downward moving particles had to be conducted in two stages. First, the Co^{60} was polarized after which the "beta and gamma counting was then started" (Wu et al, 1957, p. 1414). This configuration though only took account of the beta electrons that were moving upwards. To determine the count of the downward moving beta electrons the Co^{60} had to be repolarized by changing the magnetic field to the opposite direction. In effect this turned the Co^{60} specimen upside down with respect to the anthracene crystal so that the downward moving beta particles could be registered.

The two parts of the actual experiment are depicted at the left side of Fig. 3.4, while the right side represents what the experiment would have looked like if the apparatus could have detected the downward moving beta particles. Converting the real experiment into the counterfactual description involves flipping the upward moving beta particles of the second part of the real experiment upside down. We should note that the counterfactual configuration (the right side of Fig. 3.4) is commonly used in textbook and other secondary accounts of the experiment without comment as to its ancestry.

Assuming the correctness of the experiment result, the question now is how do we get from this experimental result to the higher-level theoretical result that parity has been violated? To see how the argument goes see Fig. 3.5 which represents in schematic form

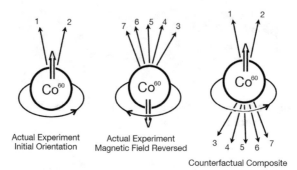

Fig. 3.4 Schematic representation of the two parts of the Wu et al. experiment and the counterfactual equivalent. Spin is represented by the double arrows pointing up or down. *Source* Authors

the essential features of the Wu experiment and its relationship with the parity transformation.[7] In order to highlight how this transformation applies we have taken the liberty of labeling the paths of the β emissions. Such identification, of course, was not available to Wu and her colleagues since only the aggregate counts for the β emissions were experimentally accessible. Nevertheless, as can be readily seen, the parity image is *not* congruent with the real-world experimental values. This because the *relative* direction of the polarization, i.e., the spin direction, of the Co^{60} changes from being the same as the minimal, originally upward β emissions to being the same as that of the maximal (originally downward) β emissions.[8]

We draw the reader's attention here to the fact that the experimental design depends on that fact that that spin (like angular momentum) is what's known as an *axial vector*. While the orientation of its rotation changes under a simple reflection transformation, the parity transformation does not alter the rotation orientation. Once it points up, it continues to do so under the parity transformation.[9]

Given then this lack of congruence between the real and the parity spaces, it follows from the theorem introduced earlier (connecting parity and congruence under parity transformation) that parity conservation has been violated.

---

[7] Note however that we have simplified insofar as we have made use of only two dimensions.

[8] It is important to keep in mind that this analysis does not assume that the β emissions are themselves polarized, only that the Co^{60} source is. It was only later determined that β emissions are in fact longitudinally polarized which is *by itself* a counterexample to parity conservation. This because in the case of a longitudinally polarized beam of β emissions, parity is violated in very much the same way that it was violated in the case of the Wu experiment, namely, that spin direction remains intact after the parity transformation while all else changes.

[9] A handy and frequently used way of stating the lack of congruence is that the pseudo-scalar $(S \cdot p)$ changes in value, where the vector S represents spin and p the momentum of the system.

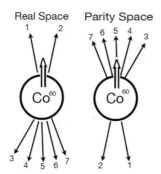

Fig. 3.5 Schematic representation of the experiment showing the "real space" result and the "parity space" counterpart. *Source* Authors

3.5   The Role Played by the Production of Gamma Rays

There is another essential aspect of the experiment, and that is that the apparatus included two gamma ray scintillation counters as shown in Fig. 3.6. As explained by Wu:

> Co^{60} nuclei can be polarized by the Rose-Gorter method in cerium magnesium (cobalt) nitrate, and *the degree of polarization detected by measuring the anisotropy of the succeeding gamma rays* (Wu et al., 1957, p. 1414, emphasis added).

The release of the gamma rays is an electromagnetic process which was known to conserve parity. Thus, while the polarization of the Co^{60} would cause variation from the initial isotropy, because parity is conserved the originally circular isotropy transforms to a symmetrical oval anisotropy.

The cryostat did eventually begin to warm (due to heat leaks and other causes) where the temperature increase had the effect of a slow reduction of polarization. Which in turn meant that the gamma ray distribution would return to an isotropic distribution. This behavior of the gamma ray distribution explains why:

> The observed gamma-ray anisotropy was used as a measure of polarization, and, effectively, temperature (Wu et al., 1957, p. 1413).

The next step was to *compare the data* for the gamma rays with that of the beta rays where the hopeful anticipation was that the declining asymmetry of the beta rays would be matched by a declining anisotropy of the gamma rays. And so it was. See Fig. 3.7. As the apparatus grew warmer the polarization decreased as shown by the fact that the gamma anisotropy values converged to the stable "warm" isotropic value. This decrease in polarization had corresponding consequences for the asymmetry of the beta emission. In fact, those corresponding consequences were exact!

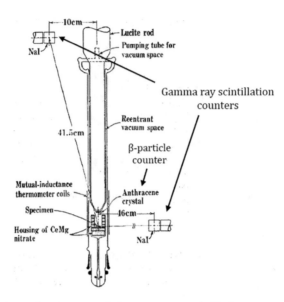

Fig. 3.6 The experimental apparatus and the gamma ray scintillation counters. *Source* Wu et al. (1957), augmented by authors

> *The disappearance of the β asymmetry coincides exactly with the time of disappearance of γ anisotropy.* The measured asymmetry indicates that the emission of electrons is preferred in the direction opposite to that of the nuclear spin. (Wu, 2008, p. 58, emphasis added)

Considering the complexities involved, this exact coincidence between the disappearance of the β asymmetry and the disappearance of γ anisotropy is astounding. Especially so when you consider the fact that the experiment had to be conducted in two distinct stages in order to separately determine the upward and downward moving β-rays. And to think that all of this depended on the quick-thinking use of "fine nylon" thread to secure the CMN housing for the specimen Co[60].

It's worth considering what exactly is the *benefit* of having this exact coincidence. For starters, the data were exactly what you'd expect if parity was not conserved *and* there were no other significant confounding causes at work. It would have been indicative of some sort of serious misalignment or other malfunction if it had not been so. Depending on the particulars, a serious misalignment would have been either difficult or even impossible to rectify. Still, that the data were what you'd expect if parity was not conserved does not mean that there were not other causes at work that were *mimicking the effect of parity non-conservation.*

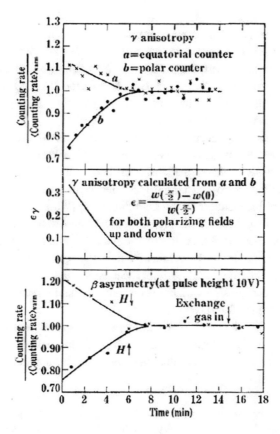

Fig. 3.7 Gamma anisotropy and beta asymmetry for different directions of polarizing field. *Source* Wu et al. (1957)

3.6 The "Systematic Checks"

I told [Lee and Yang] the effect was large and reproducible, but it must be regarded as preliminary because some *systematic checks* were not yet completed (Wu, 2008, p. 57, emphasis added).

The use of such "systematic checks," i.e. experiments performed on the experimental apparatus to determine sensitivity to possible confounding factors, is an essential part of every good experiment. Such checks are especially important when considering the pursuit worthiness of an experiment. This because such checks are often revealing as to specific avenues of improvement. A clear example of such an example of potential improvability dealt with the sensitivity of the beta ray experiments to aperture size—which was extensively and interactively discussed by Hahn, Meitner, and by Wilson. In fact, virtually every experimental report we reviewed in the preceding section contained variations in the existence and arrangement of various sorts of partitions designed to test for sensitivity to scatter. Another good example of a systematic check (which we did not earlier get to discuss) is contained in a section of Ellis and Wooster's report of their 1927 heating experiment where they dealt with the amount of gamma ray energy that managed to escape from the calorimeters (Ellis & Wooster, 1927, pp. 119–120). Here the methodology involved comparing the ability of disintegrating atoms of radium E, radium B and radium C to make such an escape.

Given the obvious importance of such "systematic checks," we'll take the opportunity to closely review the systematic checks conducted by Wu and her colleagues. There were four such sensitivity determinations. The first involved changing the direction of the (vertical) polarizing field and noting that "the warm counting rates [when there is no polarization] are independent of the field direction" (Wu et al., 1957, p. 1414). This independence argues against any significant *instrumental asymmetry*.

The second systematic check dealt with the possibility that the demagnetization field used to cool the sample might have left a "remnant magnetization" that caused the β-ray asymmetry. This confounding possibility was eliminated by noting that the observed asymmetry did not change sign with the reversal of the field direction of the (horizontal) demagnetization field (p. 1414).

A third systematic uncertainty that had to be dealt with was that there might have been a small magnetic field perpendicular to the polarizing field due to the fact that the Co^{60} crystal axis was not parallel to the polarizing field. Eliminating this possibility was somewhat more involved.

> To check *whether the beta asymmetry could be caused by such a magnetic field distortion*, we allowed a drop of CoCl solution to dry on a thin plastic disk and cemented the disk to the bottom of the same housing. In this way the cobalt nuclei should not be cooled sufficiently to produce an appreciable polarization, whereas the housing will behave as before. The large beta asymmetry was not observed (p. 1414).

The strategy here was to separate the specimen from what would otherwise be the insulating effect of the housing. By so doing, the specimen Co^{60} nuclei would no longer be in alignment and would not produce coherent (up and down) beams of β-rays. Thus, any observed beta asymmetry would have to have been produced by the secondary magnetic field caused by a lack of alignment of the Co^{60} crystal axis with the polarizing field. Since there was no such observed asymmetry there was no such "magnetic field distortion." Stated another way, what's being tested for here is the presence of any beta asymmetry in the case where there is no (or little) cooling of the specimen but where the polarizing field was in effect.

The fourth "systematic check" dealt with the case where there was cooling (by the horizontal field) of the specimen but no polarization imposed by the vertical field. Here the concern was whether "internal effects" present when the horizontal field was applied was sufficient to cause the beta asymmetry.

Furthermore, to investigate possible internal magnetic effects on the paths of the electrons as they find their way to the surface of the crystal, we prepared another source by rubbing CoCl on the surface of the cooling salt until a reasonable amount of the crystal was dissolved. We then allowed the solution to dry. No beta asymmetry was observed with this specimen (p. 1414).

If a magnetic field internal to the CMN platform upon super-cooling was sufficient to affect the CoCl specimen *in this case of cooling and non-polarization* then it would also do so in the experimental case where the Co^{60} was polarized. But since "no beta asymmetry was observed with this specimen," there was no such "internal magnetic" field.[10]

3.7   The Asymmetry Coefficient

Wu's initial response to the experiment result was only qualitative: "A large beta asymmetry was observed" (p. 1414). And while there were quantitative values for that asymmetry (1.1, 0.7), those numbers had no specific theoretical connection other than being enough to justify the judgement that parity was not conserved. Question: How was a more theoretically meaningful *quantitative measure* of the asymmetry to be *reliably* determined? How large is "large"?

The question is important because an effective quantitative measure was needed in order to connect the observed β asymmetry with ongoing theoretical developments. The answer turned on the fact that the β asymmetry could be compared with the parity invariant gamma anisotropy which served as a *parity invariant standard or benchmark*. The basic idea was to connect by means of an asymmetric coefficient the ratio of the number of up and down B-rays with a standard measure of the gamma anisotropy. In the 1957 *Physical Review* paper, Wu gives a rather truncated and consequently cryptic account of how such a quantitative comparison is to be made which yielded "*the lower limit* of the asymmetry parameter [as] approximately equal to 0.7" (p. 1414). For a more complete and understandable account see (Wu, 2008, p. 59) which among other things expressly notes that corrections taking into account scatter and the relativistic adjustment were applied to the value of the gamma anisotropy. In sum, the observed β asymmetry was equal to 0.7 (the calculated value of the asymmetry parameter) times the observed gamma anisotropy as adjusted to account for scatter and the relativistic correction.

Wu gave the following account of how this "lower limit" for the asymmetry coefficient should be understood.

---

[10] We note here that while Wu does not explicitly state that in this case the horizontal field was turned on, we think that is implicit in the argument. This is indicated by the fact that the third systematic check dealt with the case of where there was in effect no cooling but there was a vertical polarizing field. Hence, it's natural that the fourth systematic check would deal with the case where there was cooling (produced by the horizontal field) but no polarization produced by a vertical field.

In order to evaluate [the asymmetry parameter] accurately, many supplementary experiments must be carried out to determine the various correction factors. *It is estimated here only to show the large asymmetry effect....* More rigorous experimental checks are being initiated, but in view of the important implications of these observations, we report them now in the hope that they *may stimulate and encourage further experimental investigations* on the parity question in either beta or hyperon and meson decays (Wu et al., 1957, p. 1414, emphasis added).

Of significant relevance here regarding "further experimental investigations" we note that Wu had been informed by Lee and Yang that a minus 1 value for the asymmetry coefficient "implies" that "charge conjunction is also non invariant" and would be consistent with their two-component theory of the neutrino (Wu, 2008, p. 59). Thus, there was a *theoretical target* to shoot for which provided justification for the pursuit of experimental refinements that might lead to the theoretically important value of minus 1.

3.8   Replications and Improved Experimental Results

Whatever doubt there might have been on the acceptance of the "large asymmetry effect" had to have been dissipated by the fact that shortly after the Wu result was announced Garwin and Telegdi announced that they had also experimentally confirmed the nonconservation of parity this time with respect (as also suggested by Lee and Yang) to the sequential decay $\pi \to \mu \to e$ (Friedman & Telegdi, 1957; Garwin et al., 1957). Garwin's apparatus was an ingenious, but complex Rube Goldberg concoction whereas Friedman and Telegdi relied on using nuclear emulsions to separate out the decay components. These results thus served as a form of replication. In fact, as noted by Garwin and Lederman, they embarked on their experiment *after learning* that Wu had already obtained the large symmetry effect:

> Our own experiment ... was spurred by Wu's Friday-lunch report of the status of the Co-60 experiment and was performed that Friday night, 4 January 1957, to Tuesday morning 8 January, and our Letter was written that day (Garwin & Lederman, 1997, pp. 542–543).

So, for Garwin and Lederman there was no doubt as to the pursuit worthiness of some sort of replication of the Wu result.[11]

As mentioned above, there was also the question of whether further experimental investigation would lead to a convergence to the value of minus 1 for the asymmetry coefficient—which was surely a pursuit worthy venture given its theoretical importance. There was also the persistent problem of scatter to deal with as well as "the ever present stray magnetic fields in the immediate vicinity of a polarized B-source" that had to be taken into account (Chirovsky et al., 1980, p. 127). In response, Wu and her colleagues

---

[11] For some background information on the Friedman and Telegdi experiment and the publication of its results see Goudsmit (1971).

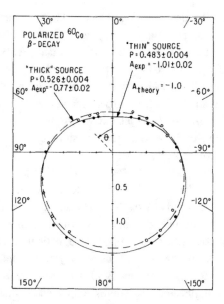

Fig. 3.8 Results of new experimental design showing "the angular distribution of β-particles from a polarized Co^{60} source," and a value for the asymmetry parameter for the "thin" source of minus 1.01 ± 0.02. *Source* Chirovsky et al. (1980)

at Columbia using a new experimental design were able to measure for the first time "the angular distribution of β-particles from a polarized Co^{60} source," and on that basis derive a value for the asymmetry parameter of minus 1.01 ± 0.02, where the error interval was "three times the statistical errors obtained from the least-squares analysis." See Fig. 3.8 This value (obtained with newly developed "thin" sources to reduce scatter) was expressly noted to be in "excellent agreement" with "the theoretical prediction of the distribution based on the two-component theory of the neutrino and the (V-A) interaction in β-decay" (Chirovsky et al., 1980, pp. 127, 130).

3.9   Conclusions: Acceptance and Pursuit

(1) In terms of acceptance and pursuit, the theta-tau puzzle was not, as emphasized by Lee and Yang, sufficient for acceptance of the proposition that parity was not conserved, but given the centrality of parity conservation, it was sufficient for experimental pursuit—despite the fact that such non-conservation would be an outlier in a field of parity conservation. Thus, being worthy of pursuit does not require that the venture be thought likely to prevail against the existing regime of accepted results, or even that it be more pursuit worthy than other nearby ventures of pursuit. There is an inescapable risk, sometimes prudential, involved in such choices.

(2) Wu's limited conclusion that there had been a "large asymmetry effect" along with the preliminary estimate of the asymmetry coefficient as minus 0.7 was sufficient—and was historically taken to be so—to justify the *acceptance* of the proposition that parity conservation was not conserved for weak force interactions. Moreover, this acceptance was in turn taken to justify the experimental *pursuit* of the consequences of parity non-conservation. Thus, an experimental result may warrant acceptance in some respects while being deemed pursuit worthy in others. As a corollary, we note that an experimental result may be deemed pursuit worthy for purposes of replication after which it may achieve acceptance.

(3) The Wu experiment was particularly noteworthy—and hence warranted acceptance or at least further pursuit—because of the intricate, and by no means guaranteed, exact coincidence of the disappearance of the β asymmetry and that of the γ anisotropy. This coincidence was especially outstanding considering that the experiment had to be conducted in two parts with contrary imposed magnetic fields. In short, the experiment was impressive because of the *internal consistency* of its interacting components. The Ellis and Wooster (1927) experiment is also noteworthy in this regard when one considers that the complicated procedures employed in measurement and analysis yielded, as an indirect result, a value of 5.1 days for the then accepted decay period of 5.0 days for radium E (Ellis & Wooster, 1927, p. 117).

(4) The Wu experiment was accompanied by an extensive and well considered set of "systematic checks" that surveyed a range of possible confounding factors and found that none were of significance, at least not when taking into account the fact that the 0.7 asymmetry parameter was a lower limit, and that there was explicit recognition of the fact that "many supplementary experiments must be carried out to determine the various correction factors" needed to justify a more restrictive measure of the asymmetry significance.

(5) Finally, the Wu experiment had the great virtue of *conceptual simplicity* which facilitated the appraisal of its internal consistency and the adequacy of the systematic checks performed. It's no wonder therefore that Wu's experiment appears on the covers of many textbooks and is *the* experiment that's cited as proof of the nonconservation of parity for weak forces.

References

Bellantoni, L. (2016). Theta and tau, two generations later. *Fermi News*. Available from: https://news.fnal.gov/2016/04/theta-tau-two-generations-later/

Bernstein, J. (1967). *A comprehensible world*. Random House.

Chirovsky, L. M., Lee, W. P., Sabbas, A. M., Groves, J. L., & Wu, C. S. (1980). Directional distributions of beta-rays emitted from polarized ^{60}Co nuclei. *Physics Letters, 94B*, 127–130.

Ellis, C. D., & Wooster, W. A. (1927). The average energy of disintegration of radium E. *Proceedings of the Royal Society (London) A, 117*, 109–123.

Feynman, R. P. (1985). *Surely you're joking, Mr. Feynman.* Norton.

Friedman, J. L., & Telegdi, V. L. (1957). Nuclear emulsion evidence for parity nonconservation in the decay chain $\pi^+ - \mu^+ - e^+$. *Physical Review, 105*, 1681–1682.

Garwin, R. L., & Lederman, L. M. (1997). History of parity violation experiment. *Nature, 386*, 542–543.

Garwin, R. L., Lederman, L. M., et al. (1957). Observation of the failure of conservation of parity and charge conjugation in meson decays: The magnetic moment of the free muon. *Physical Review, 105*, 1415–1417.

Gibson, W. M., & Pollard, B. R. (1976). *Symmetry principles in elementary particle physics.* Cambridge University Press.

Goudsmit, S. A. (1971). A reply from the Editor of Physical Review. *Adventures in Experimental Physics Gamma, 137*.

Hudson, R. P. (2016). *The reversal of the parity law in nuclear physics.* National Institute of Standards and Technology. Available from: https://www.nist.gov/pml/fall-parity/reversal-parity-law-nuclear-physics

Lee, T.D. (1985). *Letter.* A. Franklin.

Lee, T. D., & Yang, C. N. (1956). Question of parity nonconservation in weak interactions. *Physical Review, 104*, 254–258.

Wu, C. S. (2008). The discovery of the parity violation in weak interactions and its recent developments. *Lecture Notes in Physics, 746*, 43–69.

Wu, C. S., Ambler, E., Hoppes, D. D., & Hudson, R. P. (1957). Experimental test of parity nonconservation in beta decay. *Physical Review, 105*, 1413–1415.

Is There a Fifth Force?

4

In 1986 Ephraim Fischbach, Sam Aronson, and their collaborators (henceforth Fischbach) proposed a modification of Newton's Law of Universal Gravitation of the form $V = -Gmm'/r\,[1 + \alpha e^{-r/\lambda}]$, where the first term is Newton's Law and the second term is what became known as the Fifth Force,[1] α was the strength of the new force and λ was its range (Fischbach et al., 1986a). The proposal was stimulated, in part, by two tantalizing pieces of evidence: a two or three standard deviation energy dependence in the CP violating parameters in K^0_L decay and a similar size difference between measurements of G, the constant in Newton's Law, in the laboratory and in towers or mineshafts.[2]

After further investigation and analysis, Fischbach came to believe that the strength of the gravitational modification arose from a new interaction coupled to the baryon number B of a substance[3] where α, the strength of the Fifth Force, is given by

$$\alpha = -(B_1/\mu_1)(B_2/\mu_2)\xi_B$$

where $B_{1,2}$ are the baryon numbers of the interacting objects, and $\mu_{1,2}$ the corresponding masses (in units of atomic hydrogen), and where ξ_B is a defined universal constant.

---

[1] The other four forces were the strong, or nuclear, force, the electromagnetic force, the weak force, and the gravitational force.

[2] For a more details see the first-person account by Ephraim Fischbach in Franklin and Fischbach (2016, Chap. 6).

[3] For ordinary (baryonic) matter such as considered here, the baryon number is just the sum of the protons and neutrons adjusted to take account of the relative occurrence of isotopes. For the details on how to deal with isotopes and compounds see Fischbach and Talmadge (1999, 19–26).

© The Author(s), under exclusive license to Springer Nature Switzerland AG 2022 93
R. Laymon and A. Franklin, *Case Studies in Experimental Physics*,
Synthesis Lectures on Engineering, Science, and Technology,
https://doi.org/10.1007/978-3-031-12608-6_4

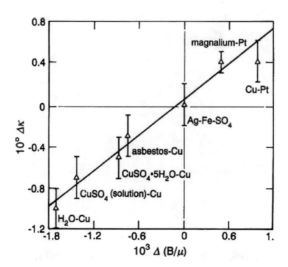

Fig. 4.1 Plot of Δk as a function of $\Delta(B/\mu)$. *Source* Fischbach et al., (1986a, 1986b)

But how was this to be experimentally tested? Here Fischbach hit upon the opportunistic and cost-effective option of reexamining the data from the original Eötvös experiment[4] with an eye to determining whether there was in fact a discoverable variation in measured gravitational attraction that varied as the product of the ratio of baryon number to mass of the interacting substances. Thus, Fischbach began with a determination of the applicable ratios of baryon number to mass, and then plotted those ratios against the Eötvös data regarding the observed angle of twist with respect to the substance comparisons made. See Fig. 4.1. Surprisingly, and shocking to many, the comparison as reported in Fischbach et al. (1986a) resulted in a statistical best fit to a linear equation of the slope to be expected based on the Fifth Force hypothesis as well as the anticipated ordering of the test substances. And where, furthermore, the range of the parameters needed to fit the data were consistent with the available geological data on G.[5] Given all this, Fischbach stated:

> Our primary conclusion is that the EPF [i.e., Eötvös] data taken *by themselves* provide fairly compelling evidence for a new intermediate range force coupling to baryon number or hypercharge. Thus even if it were to turn out that there is no connection between these data and the geophysical and [K meson] analyses which motivated our reexamination of the EPF experiment, the EPF results (if correct) *would be sufficient evidence* for the existence of a fifth force" (Fischbach et al., 1988, pp. 70–71, emphasis added).

---

[4] See Franklin and Laymon (2019, Chap. 4) for an account of this experiment and its history.

[5] For a review of these constraints see Franklin and Fischbach (2016, pp. 8–9), De Rujula (1986b), and Fischbach and Talmadge (1999, pp. 61–63).

Fischbach and his collaborators found that they could obtain reasonable fits to all the observations with α ≈ 0.01 and λ ≈ 100 m. Thus, the Fifth Force hypothesis acted as an enabling hypothesis. It told experimenters approximately the size of the effect they should observe if there was a Fifth Force.

In terms of acceptance and pursuit, it's evident that that all of this justified the further pursuit of the Fifth Force hypothesis though not its acceptance. Still, and not surprisingly, there was considerable skepticism about the apparently miraculous nature of Fischbach's new analysis of the Eötvös data—what one commentator referred to with some gentle irony as a "gorgeous bibliographical discovery" (De Rujula, 1986a, p. 761). The skepticism appeared justified when a sign error in Fischback's calculations was identified, but once rectified the situation eventually settled and an uneasy consensus developed that Fischbach had indeed managed to extract from the Eötvös data what appeared to be confirming evidence for the Fifth Force.[6]

In any case while initially promising, Fischbach's analysis of the Eötvös data suffered a significant setback when it was realized that the Eötvös torsional balance would not have been able to detect the Fifth Force unless there had been a large mass parked nearby such a hillside or otherwise substantial rock formation. To see why this is so, we begin by noting that since the Fifth Force operates only over a relatively short range, it is reasonable to begin with the assumption of a flat earth. And since the identity of inertia and gravitational mass holds to at least to within 10^{-9}, it follows under these assumptions that the Fifth Force will act perpendicularly to the earth's surface and because of that will not exert any torque on a torsional pendulum.[7] This discouraging result follows as well if one employs the more realistic assumption that "[t]o the first order, the Earth is an ellipsoid of revolution held together by gravitation but deformed by the centrifugal forces of its rotation, and the plumb line is not, in general, directed toward the center of the ellipsoid." Hence, "[h]orizontal gravitational forces... on passive gravitational masses are exactly balanced by opposing centrifugal forces... on inertial masses." Which means that "in effect there would be no horizontal 'fifth force' component, so the torsion balance would sense nothing" (Eckhardt, 1986).

But on yet further analysis Fischbach was able to show that since the Fifth Force "is of short range... the *dominant contributions* in the EPF experiment will come from local departures from the Earth's geoid." (Fischbach et al., 1986b, p. 2869, emphasis added). The solution to the question of the feasibility of an experimental test for the existence of the Fifth Force was to park one's apparatus next to a stable large hillside or equivalent rock formation.

---

[6] For a review of the sign problem and the extensive discussion involved see Fischbach and Talmadge (1999, pp. 8–9) and Franklin and Fischbach (2016, pp. 30–31, 180–181). And for the imprimatur eventually afforded to Fischbach's reanalysis of the Eötvös data see De Rujala (1986b, pp. 218–220), Fischbach et al., (1988, p. 29), and Franklin and Fischbach (2016, pp. 25–26).

[7] See Bizzeti (1987, 82–84) for elaboration along these lines.

All well and good for the pursuitworthiness of an experimental search for the Fifth Force. But here the downside was that the reliability of the Eötvös data as supporting the Fifth Force hypothesis depended on there being a rather large mass near the Eötvös apparatus. And at first glance there didn't appear to be any such mass. After making enquiries, Fischbach received information from Judit Németh regarding the site where the Eötvös experiment was performed. But while suggestive, the information upon analysis was inconclusive. Thus, as conceded by Fischbach, "neither the magnitude nor the sign of the effective hypercharge coupling can be extracted unambiguously from the EPF data without a more detailed knowledge of the local matter distribution" (Talmadge et al., 1986, pp. 237–238).

Some critics took this as confirming their skepticism. For example, and with mitigating humor, particle theorist Shelden Lee Glashow, opined that the case for the Fifth Force consisted of "unconvincing and unconfirmed kaon data, a reanalysis of the Eötvös experiment depending on the contents of the Baron's wine cellar [an allusion to the importance of local mass inhomogeneities], and a two-standard-deviation geophysical anomaly! Fischbach and his friends offer a silk purse made out of three sows' ears, and I'll not buy it." (Quoted in Schwarzschild, 1986, p. 20).

Despite all this, the *experimental search* was still determined by a significant number of experimentalists to be pursuit worthy. What explains this? Two factors come to mind. First, there was the obvious significance of the Fifth Force if it were detected. And here we note that this factor can be operative even if the experimentalists involved do not believe that the Fifth Force hypothesis is true or even likely to be true. Second, that pursuit is facilitated and made more attractive when there is the ready availability of the necessary theoretical and experimental expertise, and the possibility, given such expertise, of mounting an experimental test of the required sensitivity. We'll now review the evidence that supports the efficacy of these two factors in the case of the Fifth Force.

At the 1989 and 1990 Moriond workshops, Allan Franklin was able to discuss the interest of the experimentalists present in conducting tests to determine the existence of the Fifth Force. So, for example, Donald Eckhardt, James Faller, and Riley Newman had both an interest in pursuing such an experimental test, and moreover had already conducted or were planning experiments that could be coopted for such testing. In short, there was both theoretical interest and experimental expertise.

Eckhardt, a scientist in the Air Force Geophysical Laboratory, had been planning balloon measurements of gravity to investigate whether the lack of detailed and precise knowledge of surface gravity might account for missile accuracy problems. The arrival of the Fifth Force hypothesis provided additional motivation for such measurements and encouraged Eckhardt to bring to fruition his tower gravity experiment.

Faller had been working on gravity experiments since his days as a graduate student at Princeton in the early 1960s. When interviewed by Franklin, he was working on experiment to measure the acceleration due to gravity at the surface of the Earth that involved dropping a weight in a specially constructed chamber. He was also constructing a second chamber for use by another group. But with the arrival of the Fifth Force, Faller realized that with relatively modest modifications he could use both chambers for a Galileo-type test of whether there was a Fifth Force that resulted in a composition dependence of the acceleration.[8]

Newman had been working on the distance dependence of the inverse square law of gravity since about 1980. He was both a participant in and a coauthor of two experiments that had set very stringent limits on the deviations from the inverse square law at very short distances of the order of a few centimeters (Hoskins et al., 1985; Spero et al., 1980). Even prior to the publication of the Fifth Force paper he had made a proposal to the National Science Foundation for an experiment to investigate the possible composition dependence of the gravitational force.

For the other scientists interviewed, the investigation of this "intriguing possibility," a phrase used by both Eric Adelberger and Paul Boynton, involved changing their area of research. All the researchers remarked that the idea of testing a fundamental law of physics with a table-top experiment, that is, with a comparatively inexpensive and conceptually simple apparatus, was an important part of their motivation. Several of these investigators had worked previously on tests of other fundamental laws. David Bartlett, for example, had worked on tests of time reversal violation, a fundamental symmetry in nature, and on the distance dependence of the inverse square law of electrostatic force (Coulomb's Law). Newman had, in addition to his earlier work on gravity, also investigated whether there was a spatial asymmetry in beta decay. Adelberger noted that he had discussed the possibility of a Fifth Force test with Blayne Heckel, another member of the Eöt-Wash group, and that they had been able, within an hour, to come up with a relatively simple idea for a workable apparatus along with a preliminary analysis of background and systematic effects. They had both, in their respective work in nuclear and atomic physics, previously worked on the measurement of small effects.

It should not, however, be inferred from this that these scientists believed that the Fifth Force hypothesis itself showed promise of being true or that it was likely to be true. One might be tempted to think that experimental testing could only be deemed pursuit worthy if the underlying theory involved was a serious candidate for acceptance. But among the scientists who believed experimentation was pursuit worthy there was no unanimity on this point.

In fact, it is fair to say that the physicists Franklin spoke with were quite skeptical about the existence of a Fifth Force. Their attitudes ranged from Eckhardt's view that Fischbach and his collaborators were wrong and that he was going to demonstrate it with his tower experiment, to Newman's belief that the hypothesis had a 20–30% chance of being correct, not an overwhelmingly positive view. Adelberger remarked on the difficulty of both analyzing the data and of finding systematic effects in current experimental work and expressed doubts that they could be done well for experiments performed 75 years earlier

---

[8] Faller later remarked that the relatively modest modifications of his apparatus took six months to complete.

although he found the results of Fischbach's reanalysis of the Eötvös data very interesting. Eckhardt, despite his expressed skepticism, reported a positive Fifth Force result, at least in his early work, and Newman, who was more positive, reported an experiment that found no such effect and set some of the more stringent limits on such a force.

This lack of unanimity and its strong skeptical cast raises the question what it was about the Fifth Force that justified the pursuit of experimental evidence that would be relevant to its acceptance, further pursuit, or rejection. For the case at hand, the decision to pursue experimental testing is best understood as being an act of prudence, of hedging one's bets. And that what justified such prudence was the undeniable significance of the Fifth Force hypothesis if it were accepted. Thus, it was entirely rational even for those who were seriously skeptical as to the truth of the hypothesis to nevertheless pursue its experimental examination.

The onus was thus on the experimentalists to take advantage of the amplification possibilities of nearby mass inhomogeneities and devise experiments that were correspondingly sufficiently sensitive and ultimately decisive. And in fact, it didn't take long for the first experimental tests to be published as adjoining papers in *Physical Review Letters* (Stubbs et al., 1987; Thieberger, 1987b) and then subsequently reported at the 1989 Moriond workshop. The experiments were conducted using rather different apparatus and methods.[9]

Thieberger's experiment was noteworthy because it avoided the complication of having to take account of both gravitational and centrifugal forces since it relied only on the differential Fifth Force effect due to the Palisades cliff in New Jersey. This simplification was possible because Thieberger employed a differential accelerometer in which a copper sphere was submerged in water such that the center of mass of the copper sphere corresponded to that of the displaced water.

The Eöt-Wash group, by contrast, used a freely oscillating torsion pendulum containing two beryllium and two copper test bodies where the beryllium and copper had B/μ ratios of 0.99865 and 1.00112, respectively and thus maximized the difference in B/μ ratios. The four-body pendulum was located on the side of a hill on the University of Washington campus, which provided the local mass asymmetry needed for an observable Fifth Force effect.

This diversity meant that an agreement of result would be strongly confirming or disconfirming of the Fifth Force hypothesis because of the unlikely possibility that the different systematic uncertainties involved would all conspire to yield the same result. There was, however, disappointment on this score. The results were not in agreement. Peter Thieberger's results supported the existence of a Fifth Force (Fig. 4.2), whereas the

---

[9] These were tests of the composition dependence of the Fifth Force. There were also experimental tests of the distance dependence. These also gave discordant results, which were later resolved. For details see Franklin and Fischbach (2016, pp. 66–80). We will concentrate here on the composition dependence.

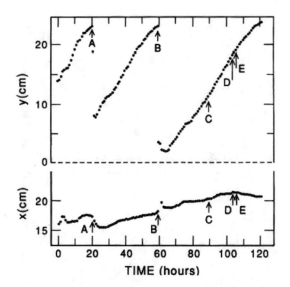

Fig. 4.2 Position of the center of the sphere as a function of time. The y axis points away from the cliff. The position of the sphere was reset at points A and B. *Source* Thieberger (1987b)

results of the Eöt-Wash group found no evidence for such a force (Fig. 4.3).[10] So, while the two experiments *considered together* had the effect of denying acceptance, they also adding urgency to the need for additional experimentation that would break the stalemate. Stated another way, while the experiments of Thieberger and the Eöt-Wash group did not rise to the status of acceptance they were sufficient when considered together—and given the potential importance of the Fifth Force—to justify further experimental pursuit.

That further pursuit was called for is shown by the many following experimental tests, including a combined half-ring torsional balance, updates of the Tower of Pisa experiment, and a series of ever more sensitive floating sphere experiments. Here though there was unanimity because none agreed with Thieberger's result.[11] See Table 4.1. In this regard the floating sphere experiments of P. G. Bizzeti and his collaborators were particularly influential because they used the same type of experimental apparatus as had Thieberger and revealed no Fifth Force effect. These experiments were especially compelling because Bizzeti conducted a *series* of experiments with corresponding improvements made in response to the experimental difficulties *specific* to the use of a

---

[10] This was the result presented at the Moriond Workshop. It used $\alpha = 0.01$ and $\lambda = 100$ m, a more realistic estimate of the Fifth Force, than that used in the Physical Review Letters paper, which used $\alpha = 0.001$. Fischbach used 0.007 and Thieberger used 0.008.

[11] For a comprehensive review see Franklin and Fischbach (2016, 49–79) and Fischbach and Talmadge (1999, 146–177).

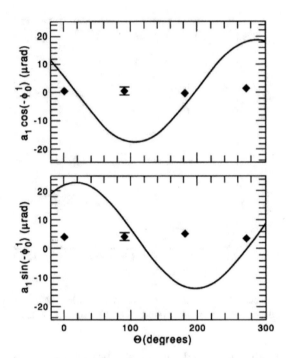

Fig. 4.3 Detection signal as a function of θ. The theoretical curves correspond to the signal expected for $\alpha = 0.01$ and $\lambda = 100$ m. *Source* Raab et al. (1987)

differential accelerometer.[12] Thus, if Bizzeti had obtained a positive result, there would have been a correspondence of result with the *type* of apparatus used. The problem then would have been to determine whether the floating-ball or torsional pendulum apparatus had the upper hand when it came to the treatment of systematic uncertainties.

Over the next two years several more results appeared in the literature. All were negative. That led Thieberger to conclude that he might have made an error:

> After the initial suggestion to perform this type of measurement close to cliffs and after the first such experiment, several other experiments now have been reported, most of them conducted with more conventional instruments. Even though the sites and the substances involved vary, effects of a magnitude expected from [Thieberger's initial paper] have not been observed.... it now seems likely that some other spurious effect may have caused the motion observed at the Palisades cliff (Thieberger, 1989, p. 2333).

At the 1990 Moriond workshop, attended by many of the researchers working on the Fifth Force, Orrin Fackler summed up the situation when he stated emphatically and without

---

[12] See Bizzeti (1987) and Bizzeti et al. (1988, 1989a, 1989b, 1990).

Table 4.1 Experimental values of α, the strength of the Fifth Force

$\lambda = 1000$ m	$\lambda = \infty$	References
$(1.1 \pm 4.3)10^{-6}$	$(1.3 \pm 5.1)10^{-10}$	Niebauer et al. (1987)
$(0.4 \pm 1.1)10^{-5}$	$(0.4 \pm 1.3)10^{-9}$	Kuroda and Mio (1990)
$(1.1 \pm 6.6)10^{-6}$	$(1.3 \pm 8.0)10^{-10}$	Kuroda and Mio (1990)
$(-0.2 \pm 1.2)10^{-5}$	$(-0.2 \pm 1.4)10^{-9}$	Kuroda and Mio (1990)
	$(-1.3 \pm 1.5)10^{-11}$	Roll et al. (1964)
	$(3.0 \pm 4.5)10^{-13}$	Braginskii and Panov (1972)
$(1.2 \pm 0.3)10^{-7}$	$-^{d}$	Boynton et al. (1987)
$(1.4 \pm 4.2)10^{-9}$	$(-0.2 \pm 1.0)10^{-11}$	Heckel (1989)
$(-5.1 \pm 5.1)10^{-9}$	$(-0.5 \pm 1.3)10^{-11}$	Heckel (1989)
$(-1.1 \pm 2.0)10^{-7}$	$(1.8 \pm 12.9)10^{-9}$	Fitch et al. (1988)
$(-7.2 \pm 7.6)10^{-4}$	$(-7.2 \pm 7.6)10^{-4}$	Speake et al. (1988)
$(-3.2 \pm 3.7)10^{-4}$	$(-3.2 \pm 3.7)10^{-4}$	Speake et al. (1988)
$(-0.7 \pm 1.4)10^{-5}$	$(-0.7 \pm 1.4)10^{-5}$	Bennett (1989)
$(0.4 \pm 3.9)10^{-6}$	$(0.4 \pm 3.9)10^{-6}$	Stubbs et al. (1989)
$(-1.4 \pm 0.9)10^{-6}$	$(-1.4 \pm 0.9)10^{-6}$	Cowsik et al. (1990)
$(1.1 \pm 1.2)10^{-6}$	$(1.1 \pm 1.2)10^{-6}$	Nelson et al. (1990)
$(-5.1 \pm 1.7)10^{-6}$	$-^{d}$	Thieberger (1987a)
$(0.0 \pm 1.1)10^{-7}$	$(0.0 \pm 1.4)10^{-9}$	Bizzet et al. (1989b)

Data taken from Adelberger et al. (1990)

qualification: "The Fifth Force is dead." No one present disagreed.[13] Less dramatically, Adelberger and other members of the Eöt-Wash group concluded:

> We have made a sensitive, systematic search for interactions mediated by ultra-low-mass scalar or vector bosons using two different detector dipoles and two different sources. We find absolutely no evidence for any new interactions ascribable to such particles. Our results break new ground over ranges from roughly 1 AU down to roughly 30 cm,[1] and are considerably more precise than any of those which claim evidence for 'new physics' (Adelberger et al., 1990, p. 3291).

The consensus then was that continuation of experimental searches for the Fifth Force was not pursuit worthy. Nor was there any attempt to determine what *exactly* had gone wrong in the Thieberger experiment such that apparently confirming data had resulted. Similarly, for Fischbach's analysis of the Eötvös data. If not explained by the now deceased

---

[13] Franklin was present at the conference.

Fifth Force, then by what? So too here, no convincing accounts have appeared.[14] In any case, the consensus that a search for such specific explanations was not pursuit worthy does serve the purpose of highlighting that there are limitations—both practical and theoretical—to what can be fully explained.

Despite what one might have expected, the demise of the Fifth Force did not cause any lack of enthusiasm among experimentalists for developing ever more accurate and expansive tests of Newtonian gravitation and in particular of the Weak Equivalence Principle (WEP). If anything, there was an upsurge in interest.[15] In short, the *experimental program* initiated by Thieberger and the Eöt-Wash group (but which had much deeper historical roots) was deemed pursuit worthy even though the initial motivation, the Fifth Force hypothesis had been deemed not pursuit worthy.[16]

There were essentially three reasons for this. First, even though the Fifth Force was no longer on the scene, the underlying theoretical basis for suspecting that there might be modifications to Newtonian gravitation was still largely intact. On the experimental side there was the accumulated expertise developed because of the extensive testing of the Fifth Force hypothesis where this expertise was at the ready for continuing investigation of such modifications. Thus, even though rejected, the Fifth Force nevertheless served to generate interest in further research regarding non-Newtonian gravitation. Fischbach took both solace and satisfaction in this positive result.

> My guess is that searches for deviations from Newtonian gravity would have had a much more difficult time becoming part of mainstream physics, had it not been for the Rencontres de Moriond and the credibility they lent to such efforts. In addition to the meetings themselves, and the opportunities they provided for interactions among the participants, the Proceedings from each meeting played an important role by collecting together many of the early experimental results and theoretical ideas (Franklin & Fischbach, 2016, p. 194).[17]

---

[14] See Fischbach and Talmadge (1999, pp. 213–214), Franklin and Fischbach (2016, pp. 204–208), and for the curious and unexplained correlation of the baryon-to-mass ratios and charge-to-mass ratios see Hall et al. (1991).

[15] For details on these experiments see Franklin and Fischbach (2016, Chap. 5).

[16] For an extensive review and analysis of one very distinguished line of historical origin see Franklin and Laymon (2019, Chaps. 2–4).

[17] In 2021, however, Fischbach and his collaborators proposed a revival of the search for the Fifth Force with recommended modifications. "Indications of a possible composition-dependent fifth force, based on a reanalysis of the Eötvös experiment, have not been supported by a number of modern experiments. Here, we argue that searching for a composition-dependent fifth force necessarily requires data from experiments in which the acceleration differences of three or more independent pairs of test samples of varying composition are determined. We suggest that a new round of fifth-force experiments is called for, in each of which three or more different pairs of samples are compared" (Fischbach et al., 2021).

Second, there was the continuing fundamental importance of the WEP for the General Theory of Relativity, and, in particular, for attempts to integrate it with the standard model of particle physics.

> The equivalence of gravitational mass and inertial mass is assumed as one of the most fundamental principles in nature. Practically every theoretical attempt to connect general relativity to the standard model allows for a violation of the equivalence principle. Equivalence-principle tests are therefore important tests of unification scale physics far beyond the reach of traditional particle physics experiments. The puzzling discoveries of dark matter and dark energy provide strong motivation to extend tests of the equivalence principle to the highest precision possible (Schlamminger et al., 2008, 041101-1, footnote omitted).

Third, it was realized that experimental examinations of WEP could be understood and used both as *tests* of WEP and as *methods of discovery* for new forms of non-Newtonian gravitation.

> The universality of free fall (UFF) asserts that a point test body, shielded from all known interactions except gravity, has an acceleration that depends only on its location. The UFF is closely related to the gravitational equivalence principle, which requires an exact equality between gravitational mass m_g and inertial mass m_i and therefore the universality of gravitational acceleration. Experimental tests of the UFF have *two aspects* – they can be viewed as *tests* of the equivalence principle or *as probes for new interactions* that violate the UFF (Su et al., 1994, p. 3614, emphasis added).

At this point we'll draw this chapter to a close but not before noting this central fact about the search for the Fifth Force. Namely, that the *experimental program* that arose in response in order to conduct this search took on a *collective life of its own* in the following respects. First, as we have noted, the experimental program was deemed pursuit worthy even by those who had serious doubts about the Fifth Force. You didn't have to be a Fifth Force proponent to be in favor of pursuing the experimental search for the Fifth Force. Second, the pursuitworthiness of the experimental program continued despite that fact that the Fifth Force had been declared dead. This because, as we have seen, it was readily co-opted as both a test and a probe for other theoretical purposes. Finally, and analogously to the developmental history of a *theoretical* research program, some of the individual constituents of the experimental program garnered acceptance (such as those of the Eöt-Wash group) while others (such as Thieberger's experiment) were denied acceptance and moreover deemed not worthy of further pursuit.

References

Adelberger, E. G., Stubbs, C. W., Heckel, B. R., Su, Y., Swanson, H. E., Smith, G., Gundlach, J. H., & Rogers, W. F. (1990). Testing the equivalence principle in the field of the Earth: Particle physics at masses below 1 μeV? *Physical Review D, 42*, 3267–3292.

Bennett, W. R. (1989). Modulated-source Eötvös experiment at little goose lock. *Physical Review Letters, 62*, 365–368.

Bizzeti, P. G. (1987). Forces on a floating body: Another way to search for long-range, B-dependent interactions. In O. Fackler, & T. T. Van Moriond (Eds.), *Moriond Workshop* (pp. 591–598). Editions Frontieres.

Bizzeti, P. G., Bizzeti-Sona, A. M., Fazzini, T., Perego, A., & Taccetti, N. (1988). New search for the 'fifth force' with the floating-body method: Status of the Vallambrosa experiment. In O. Fackler, & T. V. Thanh (Eds.), *5th Force Neutrino Physics: Eighth Moriond Workshop* (pp. 511–524). Editions Frontieres.

Bizzeti, P. G., Bizzeti-Sona, A. M., Perego, A., & Taccetti, N. (1989a). Search for a composition dependent fifth force: Results of the Vallambrosa experiemnt. In *Moriond Workshop, Monriond*. Editions Frontieres.

Bizzeti, P. G., Bizzeti-Sona, A. M., Fazzini, T., Perego, A., & Taccetti, N. (1989b). Search for a composition-dependent fifth force. *Physical Review Letters, 62*, 2901–2904.

Bizzeti, P. G., Bizzeti-Sona, A. M., Perego, A., & Taccetti, N. (1990). Recent tests of the Vallambrosa experiment. In O. Fackler, & J. Tran Than Van (Eds.), *Moriond Workshop* (pp. 263–268). Editions Frontieres.

Boynton, P. E., Crosby, D., & Newman, R. D. (1987). Search for an intermediate-range composition-dependent force. *Physical Review Letters, 59*, 1385–1389.

Braginskii, V. B., & Panov, V. I. (1972). Verification of the equivalence of inertial and gravitational mass. *JETP Letters, 34*, 463–466.

Cowsik, R., Krishnan, N., Tandon, S. N., & Unnikrishnan, S. (1990). Strength of intermediate-range forces coupling to isospin. *Physical Review Letters, 64*, 336–339.

De Rujula, A. (1986a). Are there more than four. *Nature, 323*, 760–761.

De Rujula, A. (1986b). On weaker forces than gravity. *Physics Letters B, 180*, 213–220.

Eckhardt, D. H. (1986). Comment on 'Reanalysis of the Eötvös experiment.' *Physical Review Letters, 57*, 2868.

Fischbach, E., Sudarsky, D., Szafer, A., Talmadge, C., & Aronson, S. H. (1986a). Reanalysis of the Eotvos experiment. *Physical Review Letters, 56*, 3–6.

Fischbach, E., Aronson, S. H., & Talmadge, C. (1986b). Response to Eckhardt. *Physical Review Letters, 57*, 2869.

Fischbach, E., Sudarsky, D., Szafer, A., Talmadge, C., & Aronson, S. H. (1988). Long-range forces and the Eötvös experiment. *Annals of Physics, 183*, 1–89.

Fischbach, E., & Talmadge, C. L. (1999). *The search for non-Newtonian gravity*. Springer.

Fischbach, E., Gruenwald, J. T., Krause, D. E., McDuffie, M. H., Mueterthies, M. J., & Scarlett, C. Y. (2021). Significance of composition-dependent effects in fifth-force searches. *Physics Letters A, 399*, 1276300.

Fitch, V. L., Isaila, M. V., & Palmer, M. A. (1988). Limits on the existence of a material-dependent intermediate-range force. *Physical Review Letters, 60*, 1801–1804.

Franklin, A., & Fischbach, E. (2016). *The rise and fall of the fifth force*. Springer.

Franklin, A., & Laymon, R. (2019). *Measuring nothing, repeatedly*. Morgan and Claypool.

Hall, A. M., Armbruster, H., Fischbach, E., & Talmadge, C. (1991). Is the Eotvos experiment sensitive to spin? In W.-Y. P. Hwang, et al. (Eds.), *Progress in High Energy Physics* (pp. 325–339). North Holland.

Heckel, B. R. (1989). Experimental bounds on interactions mediated by ultralow-mass Bosons. *Physical Review Letters, 63*, 2705–2708.

Hoskins, J. K., Newman, R. D., Spero, R., & Schultz, J. (1985). Experimental tests of the gravitational inverse-square law for mass separations from 2 to 105 cm. *Physical Review D, 32*, 3084–3095.

Kuroda, K., & Mio, N. (1990). Limits on a possible composition-dependent force by a Galilean experiment. *Physical Review D, 42*, 3903–3907.

Nelson, P. G., Graham, D. M., & Newman, R. D. (1990). Search for an intermediate-range composition-dependent force coupling to N-Z. *Physical Review D, 42*, 963–976.

Niebauer, T. M., McHugh, M. P., & Faller, J. E. (1987). Galilean test for the fifth force. *Physical Review Letters, 59*, 609–612.

Raab, F. J., Adelberger, E. G., Gundlach, J., & Stubbs, C. (1987). Search for an intermediate-range interaction: Results of the Eot-Wash I experiment. In O. Fackler, & J. Tran Than Van (Eds.), *New and Exotic Phenomena: Seventh Moriond Workshop*. Editions Frontieres.

Roll, P. G., Krotkov, R., & Dicke, R. H. (1964). The equivalence of inertial and passive gravitational mass. *Annals of Physics, 26*, 442–517.

Schlamminger, S., Choi, K. Y., Wagner, T. A., Gundlach, J. H., & Adelberger, E. G. (2008). Test of the equivalence principle using a rotating torsion balance. *Physical Review Letters, 100*, 041101-041101–041101-041104.

Schwarzschild, B. (1986). Reanalysis of old Eötvös data suggests 5th force ... to some. *Physics Today, 39*(Lusignoli), 17–20.

Speake, C. C., & Quinn, T. J. (1988). Search for a short-range, isospin coupling component of the Fifth Force with the use of a beam balance. *Physical Review Letters, 61*, 1340–1343.

Spero, R., Hoskins, J. K., Newman, R., Pellam, J., & Schultz, J. (1980). Tests of the gravitational inverse-square law at laboratory distances. *Physical Review Letters, 44*, 1645–1648.

Stubbs, C. W., Adelberger, E. G., Raab, F. J., Gundlach, J. H., Heckel, B. R., McMurry, K. D., Swanson, H. E., & Watanabe, R. (1987). Search for an intermediate-range interaction. *Physical Review Letters, 58*(11), 1070.

Stubbs, C. W. (1989). Limits on composition-dependent interactions using a laboratory source: Is there a "Fifth Force" coupled to isospin? *Physical Review Letters, 62*, 609–612.

Su, Y., Heckel, B. R., Adelberger, E. G., Gundlach, J. H., Harris, M., Smith, G. L., & Swanson, H. E. (1994). New tests of the universality of free fall. *Physical Review D, 50*(6), 3614.

Talmadge, C., Aronson, S. H., & Fischbach, E. (1986). Effects of local mass anomalies in Eotvos type experiements. In T. T. Van (Ed.), *Moriond Workshop*. Editions Frontieres.

Thieberger, P. (1987a). *Search for a new force*. Brookhaven National Laboratory.

Thieberger, P. (1987b). Search for a substance-dependent force with a new differential accelerometer. *Physical Review Letters, 58*, 1066–1069.

Thieberger, P. (1989). Thieberger replies. *Physical Review Letters, 62*, 2333.

The Search for Neutrinoless Double Beta Decay

5

One of the still unanswered questions concerning the neutrino is whether the neutrino is its own antiparticle (Majorana) or whether the neutrino and antineutrino are different particles (Dirac). Physicists have usually assumed that there is a distinction between the neutrino and its antiparticle, the antineutrino. This is a natural consequence of Dirac's theory in which there are two solutions to the Dirac equation, one with negative energy and one with positive energy. Electrons, for example, correspond to the case where the negative energy states are all filled. The electron antiparticle, the positron, is created when a photon of sufficient energy ($\geq 2m_ec^2$) raises an electron from a negative energy state, through a gap of twice the rest mass of the electron, to a positive energy state. The "hole" left in the negative energy "sea" behaves in the same way as a positively charged electron and is in effect the positron.[1] For charged particles, such as the electron, it was clear that the particle and the antiparticle had opposite charges. For neutral particles the consequences of the Dirac equation were less obvious.

Thus, there is in the Dirac account a striking asymmetry between the treatment of electrons and their antiparticle positron. Roughly speaking while electrons have a substantive and independent existence, positrons exist only as holes in the Dirac sea. Once however the mathematical formalism takes over the observational consequences are symmetrical. Ettore Majorana found this account overly complicated and in 1937 published a theory of electrons and positrons that was symmetrical from the start and did not require compensating mathematics to yield a symmetrical outcome.[2] As summarized by Majorana:

---

[1] For a brief account of Dirac's development of his theory of the electron "sea" and its connection with the discovery of the positron see Franklin and Laymon (2021, pp. 49–53).

[2] Majorana's article (Majorana, 1937) has been reprinted along with a translation and commentary by L. Maiani in Majorana (2006). All page references herein are to this translation.

R. Laymon and A. Franklin, *Case Studies in Experimental Physics*,
Synthesis Lectures on Engineering, Science, and Technology,
https://doi.org/10.1007/978-3-031-12608-6_5

The interpretation of the so-called "negative energy states" proposed by Dirac leads, as is well known, to a substantially symmetric description of electrons and positrons. ... The prescriptions needed to cast the theory into a symmetric form ... are however not entirely satisfactory, because one always starts from an asymmetric form or because symmetric results are obtained only after one applies ... procedures ... that one should possibly avoid. For these reasons, we have attempted a new approach, which leads more directly to the desired result (Majorana, 2006, pp. 218–219).

So, we have a situation where by design the newcomer exactly matches the observational successes of the established theory.[3] Insofar as Majorana's account is symmetric ab initio, it is without doubt an aesthetic triumph. But does that mean that Majorana's theory is worthy of pursuit? In this regard, we note that Majorana emphasized the possibilities inherent in his account for certain extensions that went beyond that of simply matching Dirac's account in terms of observational consequences.

In the case of electrons and positrons, we may anticipate only a formal progress; but we consider it important, *for possible extensions by analogy*, that the very notion of negative energy states can be avoided. We shall see, in fact, that it is perfectly, and most naturally, possible to formulate a theory of elementary neutral particles which do not have negative (energy) states (p. 219, emphasis added).

By "possible extensions by analogy" Majorana had in mind the fact that his theory "can be obviously modified" so that "there is now no need to assume the existence of antineutrons or antineutrinos." Here he had in mind Giancarlo Wick's theory of "positive β-ray emission" which involved the production of neutrinos.[4] Continuing in this vein Majorana pressed the advantages of his theory with respect to neutral particles since it,

allows not only to cast the electron-positron theory into a symmetric form, but also to construct an essentially new theory for particles not endowed with an electric charge (neutrons and the hypothetical neutrinos). *Even though it is perhaps not yet possible to ask experiments to decide between the new theory and a simple extension of the Dirac equations to neutral particles, one should keep in mind that the new theory introduces a smaller number of hypothetical entities, in this yet unexplored field* (pp. 219–220, emphasis added).

One has to ask, however, whether any of this constitutes a demonstration of promise (and thus pursuit worthiness) or merely of possibilities for theory extensions with certain satisfying symmetries. And whether, perhaps, the latter is sufficient to indicate promise.

---

[3] In this regard this case stands in contrast to Laudan's motivating examples (discussed in our Introduction) of the battle for acceptance between the Galilean research tradition and the cosmological tradition of Aristotle and Ptolemy, and that between Dalton's early atomic doctrine and the much older elective affinity chemistry.

[4] See Majorana (2006, p. 223 and note 5). For commentary on this point see Majorana (2006), Pontecorvo (1982, C8-226–C8-230), and Maiani's comments at Majorana (2006, pp. 232–233).

Giulio Racah was able to weigh the scales in favor of pursuitworthiness when he gave a formal demonstration that there was a possible beta decay process, now known as neutrinoless double beta decay (0νββ), that could be used to determine whether neutrinos were or were not their own antiparticles (Racah, 1937), with translation in Racah (1982). As summarized by Racah "thus we see that the theory of E. Majorana is not of a mere formal interest but leads us to certain physical consequences fundamentally different from these derived from Fermi's [and Dirac's] theory" (Racah, 1982, p. 39).

Continuing this pursuit of the theoretical contours for 0νββ, in 1939 Wendell Furry calculated that the rate of decay for 0νββ would be significantly higher than that for a two-neutrino beta decay mode (2νββ) and suggested that this difference might make feasible the experimental detection of 0νββ (Furry, 1939). (Later work showed that Furry's result was incorrect. Neutrinoless double beta decay is less probable than ordinary double beta decay.) In 1952 H. Primakoff calculated the electron–electron angular correlations and electron energy spectra for both 0νββ and the less controversial 2νββ (Primakoff, 1951).

Before considering how an experimental test for 0νββ was actualized we need, by way of background, to briefly discuss 2νββ. The possibility of 2νββ was first proposed by Maria Goeppert-Mayer who derived an expression for the decay rate and estimated a half-life of $\sim 10^{17}$ y for a decay with the emission of two electrons and two anti-neutrinos (Goeppert-Mayer, 1935). Here a radioactive nucleus with charge Z and mass A decays into a daughter nucleus with charge $Z \pm 2$ and mass A, with the emission of two electrons (positrons) and two as per Dirac antineutrinos, or neutrinos as per Majorana. (For example, $^{76}Ge \rightarrow {}^{76}Se + 2e^- + 2v_e^{bar}$, where the last term is a measure of released energy.) Although an earlier geochemical experiment (Inghram & Reynolds, 1950) had claimed that 2νββ had been detected,[5] the first direct experimental detection of 2νββ was not made until 1987 (Elliott et al., 1987).[6] But this experiment (and its many replications) could not, of course, distinguish between Majorana and Dirac given Majorana's deliberate attempt to capture what Dirac had to offer in such cases.

The decay process exhibited in 0νββ is similar to that of 2νββ except that no neutrinos are omitted. This is significant since such a decay mode would violate lepton conservation and is accordingly forbidden by the Standard Model (SM).[7] Dirac's theory by contrast

---

[5] A geochemical experiment "is based on analyzing an ancient mineral containing a double-β isotope with the aim of extracting and counting the number of daughter atoms of the double-β transition accumulated over long geological times." For more details see Saakyan (2013).

[6] For a comprehensive account of the difficult history and development of this ultimately successful experiment by one of the participants see Moe (2014).

[7] There are three leptons, the electron, the muon, and the tau lepton, each with its own antiparticle, and its own neutrino (antineutrino). It has been found that in all interactions so far detected lepton family number is conserved. For example, beta decay, a process in which a neutron decays into a proton, and electron, and an electron antineutrino ($n \rightarrow p + e + v_e^{bar}$). The neutron and proton have electron family number 0, whereas the electron is $+1$ and the electron antineutrino is -1. Thus, electron family number is conserved. This is also true for the muon and tau families. For double beta decay $(A, Z) \rightarrow (A, Z + 2) + 2e^- + 2v^{bar}$ electron family number is conserved. For neutrinoless

does not allow for $0\nu\beta\beta$ and thus retains consistency with SM. So, the possibility of such a crucial experiment would suggest the pursuit worthiness of such an experimental venture. More than just suggest because lepton conservation is an "accidental global symmetry in the SM" (Dolinski et al., 2019), and because of this there is an almost irresistible invitation and motivation to pursuit in the hope of discovering the underlying symmetries that are controlling. Thus, it was not surprising that there was consensus on the pursuitworthiness of an experimental search for $0\nu\beta\beta$. The case for pursuit was succinctly summarized by Klapdor-Kleingrothaus and Krivosheina:

> Nuclear double beta decay provides an extraordinarily broad potential to search for beyond standard model physics. Its occurrence has enormous consequences: it means that total lepton number is not conserved. Second, it proves that the neutrino is a Majorana particle. Furthermore, it can provide, under some assumptions, an absolute scale of the neutrino mass, and yields sharp restrictions for SUSY models, leptoquarks, compositeness, left-right symmetric models, tests of special relativity and equivalence principle in the neutrino sector, and others (Klapdor-Kleingrothaus & Krivosheina, 2006, p. 1547).[8]

The experiment to observe neutrinoless double beta decay was, in principle, straightforward. The sum of the energies of the two decay electrons had a definite value determined by the masses of the decaying nucleus and the daughter nucleus. A peak above background in the sum-energy distribution at the appropriate energy would indicate the presence of neutrinoless decay. In other words, and good news for the possibility of an experimental test, the energy spectrum for $0\nu\beta\beta$ conveniently differs from that of $2\nu\beta\beta$ since $0\nu\beta\beta$ should have a definite value whereas the energy spectrum for electrons from $2\nu\beta\beta$ will be continuous.[9]

The estimation of background is crucial in such an experiment. As one can see in Fig. 5.1 there are typically several observed peaks. If these are due to identifiable phenomena such as nuclear γ ray emission, or other identifiable processes then they should be subtracted resulting in a lower background. If, however, they are merely statistical fluctuations then they should be included in the background estimate resulting in a higher background. As we shall see there was serious disagreement on how this distinction was

---

double beta decay $(A, Z) \rightarrow (A, Z + 2) + 2e^{-}$ the initial state has electron family number 0, whereas the final state has electron family number 2. Thus, lepton number would not be conserved in such a decay process.

[8] Regarding still unanswered questions concerning neutrino masses, we note that the observed neutrinos, electron, muon, and tau are described in terms of three unknown quantities, m_1, m_2, and m_3. "It is then common to distinguish three mass patterns: Normal hierarchy, where $m_1 < m_2 < m_3$, inverted hierarchy, where $m_3 < m_1 < m_2$ and the quasidegenerate spectrum, where the differences between the masses are small with respect to their absolute values (Giuliani & Poves, 2012, p. 857016-2)." If one can determine or set limits on the Majorana neutrino mass this would determine which of these options is correct.

[9] See Saakyan (2013, pp. 507–10) for a comprehensive explanation and analysis of the determination of the energy spectrum.

Fig. 5.1 Sum energy spectrum for double β decay of ^{76}Ge. If neutrinoless double beta decay had been observed there should have been a peak at 2039 keV. No evidence for such a peak is seen. *Source* Baudis et al. (1997)

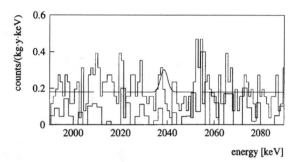

to be effectuated. On a more positive note because the decay electrons are emitted at the same time and place (Single Site Events) whereas many of the background processes occur at different points in space, experimenters were able to use what's known as pulse shape analysis to help separate signal from background events.

Before proceeding we need to note that in addition to the difficulties associated with the determination of experimental background there were also a number of theoretical problems associated with beta decay that added further uncertainty to any experimental endeavor. These involved the fundamental nature of the decay process itself and more particularly the determination of what was known as the "nuclear matrix" which was needed to compare experimental results involving different sources as well as the determination of the half-life of the decay process.[10] But rather than pursue such theoretical developments, and in keeping with our focus on the acceptance and pursuit worthiness of experimental results, we'll turn our attention to some of the initial attempts to detect the existence of 0νββ.

Experimental results on neutrinoless double beta decay began to be reported in the late 1990s.[11] Of particular interest is the work of the Heidelberg-Moscow collaboration. The group used 86% enriched ^{76}Ge crystals and reported that:

> In both sets of data [with and without single-site identification as shown in Fig. 5.1] we do not see any indication for a peak at the decay energy (Baudis et al., 1997, p. 222).

Their observed half-life limit of $T^{0\nu}_{1/2} > 1.1 \times 10^{25}$ yr (90% C.L.) set an upper limit of 0.46 eV on the mass of the Majorana neutrino.[12]

In 2001 the Heidelberg-Moscow collaboration reported results from a larger data set from their continuing experiment. For the full data set with no pulse shape analysis, they concluded:

---

[10] For an extensive review of these theoretical issues see Dolinski et al. (2019, pp. 221–233).

[11] For a more complete account of these experiments and in particular the back and forth between Klapdor-Kleingrothaus et al. and their critics see Franklin (2018, pp. 11-4–11-35).

[12] For a brief review of the theoretical determination (based on experimental values) of 0νββ half-life see Dolinski et al. (2019, p. 223).

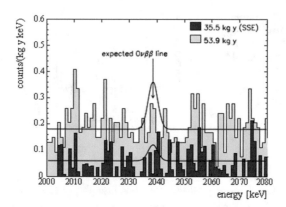

Fig. 5.2 Sum energy spectrum of all five detectors with 53.9 kg y and SSE [single site events] spectrum with 35.5 kg y in the region of interest for the $0\nu\beta\beta$ decay. The curves correspond to the excluded signals with $T^{0\nu}_{1/2} \geq 1.3 \times 10^{25}$ y (90% C.L.) and $T^{0\nu}_{1/2} \geq 1.9 \times 10^{25}$ y (90% C.L.), respectively. *Source* Klapdor-Kleingrothaus et al. (2001a)

> The expected background in the $0\nu\beta\beta$ region is estimated from the energy interval 2000-2080 keV.... The expected background in the 3σ peak interval, centered on 2038.56 keV [The expected sum energy for neutrinoless double beta decay] interpolated from the adjacent energy regions is (110.3 ± 3.9) events. The number of measured events in the same peak is 112 (Klapdor-Kleingrothaus et al., 2001a, p. 151).

Using the pulse shape analysis they expected 20.4 ± 1.6 background events and they observed 21 events. Neither data set showed evidence for a peak at the expected energy (Fig. 5.2). For the full data set they set limits of $T^{0\nu}_{1/2} > 1.3 \times 10^{25}$ y and on a Majorana neutrino mass of less than 0.42 eV, at the 90% confidence level. Using the pulse shape analysis, they found for Single Site events (SSE) limits of $T^{0\nu}_{1/2} > 1.9 \times 10^{25}$ y and 0.35 eV. The group concluded that, "We see in none of the two data sets an indication for a peak at the Q-value of 2038.56 keV for the $0\nu\beta\beta$ decay" (p. 150).

Later in 2001 three members of the Heidelberg-Moscow collaboration, plus H. Harney, claimed that they had found evidence for ^{76}Ge neutrinoless double beta decay (Klapdor-Kleingrothaus et al., 2001b). This unexpected turnabout disagreed not only with their 1997 result, but also with the group's recent 2001 report.[13] Using Bayesian analysis, which had not been used in the earlier 2001 analysis, they found a 2.2 standard-deviation effect, with a best value for the decay half-life of 1.5×10^{25} y. Their experimental result is shown in

---

[13] This is not a unique occurrence. In 1995 different members of the LSND collaboration published discordant results on neutrino oscillations in adjoining papers in *Physical Review Letters*. For a discussion of this episode see Franklin (2001, 301–311).

Fig. 5.3 Sum energy spectrum of the ^{76}Ge detectors over the period August 1990 to May 2000, 46.502 kg y. The curve results from Bayesian inference. It corresponds to a half-life $T_{1/2}^{0\nu} = (0.75 - 18.33) \times 10^{25}$ y (95% C.L.). *Source* Klapdor-Kleingrothaus et al. (2001b)

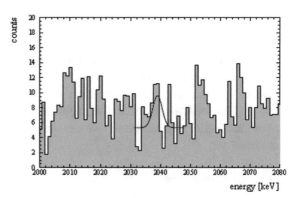

Fig. 5.3. There is no obvious peak seen at 2039 keV, the sum energy expected for ^{76}Ge neutrinoless double beta decay. That peak appeared only in the Bayesian analysis.[14]

Not everyone agreed with this claim of a positive result and criticism was not long in coming.[15] In 2002, Aalseth et al. did not criticize the "high quality data" on which the positive claim was based, but rather the analysis procedures that produced it.

There are three unidentified peaks in the region of analysis that have greater significance than the 2039-keV peak. There is no discussion of the origin of these peaks. There is no discussion of how sensitive the conclusions are to different mathematical models. There is a previous Heidelberg-Moscow publication that gives a lower limit of 1.9×10^{25} y (90% confidence level). This is in conflict with the "best value" of the new [Klapdor-Kleingrothaus et al., 2001b] paper of 1.5×10^{25} y. This indicates a dependence of the results on the analysis model and the background evaluation (Aalseth et al., 2002, p. 1476).

A similar criticism was offered by Zdesenko et al. (2002). They performed both a simple statistical analysis and a fit to both the 2001 Heidelberg-Moscow data and to a combination of that data with additional data from another experimental collaboration known as IGEX.

The claim of discovery of the neutrinoless double beta ($0\nu2\beta$) decay of ^{76}Ge (Klapdor-Kleingrothaus et al., 2001b) is considered critically and firm conclusion about, at least, prematurity of such a claim is derived on the basis of a simple statistical analysis of the measured spectra (p. 209).

Zdesenko and his collaborators found that while there was a small effect, it was only seen when the energy range chosen for the analysis is 2032–2047 keV. "Thus, one can

---

[14] Bayesian analysis employs a statistical analogue of Bayes' theorem whereby initially held beliefs as to probability are updated on the basis of new evidence. A crucial problem here is to determine the sensitivity of the final results with respect to such prior assignments of probability.

[15] The Heidelberg-Moscow 2001b paper aroused considerable interest. Klapdor-Kleingrothaus (2002) listed 28 papers on the subject.

conclude that the (2–3) σ 'effect' is only due to the unique (and incorrect) choice of the fitting interval" (p. 210).

The Heidelberg-Moscow collaboration continued to take data and evidently undeterred by the criticism Klapdor-Kleingrothaus and collaborators published a paper in 2004 reporting a neutrinoless double beta decay effect with an *increased statistical significance of 4.2σ* (Klapdor-Kleingrothaus et al., 2004a).[16] This was based on a larger data set and a more sophisticated pulse shape analysis. In reviewing this experiment Avignone et al. concluded that:

> At least nominally, the experiment had a 4σ discovery potential, and cannot be dismissed out of hand. Since this analysis does not account for statistical fluctuations, background structure, or systematic uncertainties, the actual confidence level could be significantly different. But the only certain way to confirm or refute the claim is with additional experimentation, preferably in ^{76}Ge (Avignone, 2008, p. 484).

Confidence in that "discovery potential" was not helped when once again there was discord within the Heidelberg-Moscow collaboration when a different subgroup reported a negative result when analyzing the same data. Bakalyarov and collaborators (2005) demonstrated that their background model fit their observed data and concluded that there was no evidence for neutrinoless double beta decay. They estimated that the lifetime was $T^{0\nu}_{1/2} \geq 1.55 \times 10^{25}$ y.

The CUORICINO collaboration presented yet another negative result using a different nucleus, ^{130}Te, and a different experimental technique. Their detector consisted of 62 TeO bolometers cooled to ~ 8 mK and located in the Grand Sasso Underground Laboratory which reduced the background due to cosmic rays. Once again 0νββ had escaped detection.

> Only the discovery of peaks at the energies expected for 0νββ decay, in two or more candidate nuclei, would definitely prove its existence. *No positive evidence has been reported so far*, with the exception of the claimed discovery of the decay of ^{76}Ge reported by a subset of the Heidelberg-Moscow collaboration (Arnaboldi et al., 2005, p. 142501-1 emphasis added).

They calculated a limit for the lifetime $T^{0\nu}_{1/2} \geq 1.8 \times 10^{24}$ years and an upper limit for the neutrino mass of between 0.2 and 1.1 eV, depending on the nuclear matrix element employed. With respect to the 2004 discovery claim made by the Heidelberg-Moscow collaboration they expressly noted that "[t]he range of bounds, 0.2–1.1 eV, partially covers the range of values 0.1–0.9 eV corresponding to the evidence claimed by Klapdor-Kleingrothaus et al. (p. 142501-1)."

At first glance it seems obvious that evidence from different experiments and in particular as here those performed on different nuclei, provides more support for a hypothesis

---

[16] A more detailed account of data acquisition and analysis appeared in Klapdor-Kleingrothaus et al. (2004b).

or an experimental result than does a replication of the same the original experiment using the same nucleus.[17] But, as against this, the uncertainty in the nuclear matrix elements argues for improved experiments performed on ^{76}Ge. Before getting to these better experiments, we note that in 2006 Klapdor-Kleingrothaus and Krivosheina published what they stated was "a final detailed discussion of the results of the analysis of the Heidelberg-Moscow experiment for the measuring period 1995–2003."

> We show that we have now two projections of events with almost no background, which both prove the existence of a line at $Q_{\beta\beta}$ [the expected sum energy] which we identify as $0\nu\beta\beta$ signal. This signal seen in the pulse shape analyzed spectra *has a confidence level of ~6σ* (Klapdor-Kleingrothaus & Krivosheina, 2006, emphasis added).

They calculated a lifetime $T^{0\nu}_{1/2} = \left(2.23^{+0.44}_{-0.31}\right) \times 10^{25}$ y. This also gave a Majorana neutrino mass of $m_\nu = 0.32 \pm 0.03$ eV.

With respect to the claimed confidence level of approximately 6σ we note that by this time there had developed an informal consensus in high energy physics that a confidence level of at least 5σ was required as a necessary condition to qualify as a discovery.[18] Thus, if all else was in order, then the experimental result should be accepted as a genuine discovery, in this case, of $0\nu\beta\beta$. But as we have seen, Klapdor-Kleingrothaus et al. were already under a cloud with respect to their earlier analysis of their data, and moreover their latest data analysis, even if pumped up to 6σ, continued to be a distinct outlier in a substantial sea of otherwise negative experimental results. All of which suggested that not all else was in order.

In 2008 a second generation of experimental results on neutrinoless double beta decay began to appear. These involved larger data sets, more sophisticated analysis procedures, different nuclei, and different experimental techniques.[19] None of these experiments, with the exception of those reported by a few members of the Heidelberg-Moscow collaboration, had found evidence for neutrinoless double beta decay, but, with the usual caveats concerning the uncertainties concerning the calculation of nuclear matrix elements, they were able to set upper limits on the mass of the proposed Majorana neutrino. Some of these limits overlapped with the Heidelberg-Moscow results, but other experimental limits were higher. We will present only two examples.

One stringent test of the Heidelberg-Moscow positive result was provided by the GERDA (Germanium Detector Array) collaboration (Agostini et al., 2013). This experiment used ^{76}Ge detectors from both the Heidelberg-Moscow experiment and the IGEX experiment. As mentioned earlier, other experimental results on other nuclei could compare their results to those of the Heidelberg-Moscow collaboration only by using nuclear

---

[17] For a Bayesian account see Franklin and Howson (1984).

[18] For a review of the evolution of the 5σ standard see Franklin (2013, ix–lv).

[19] See Barabash (2011) for a comprehensive review of many of the early twenty-first century experiments that searched for double beta decay, both with and without neutrinos. By 2011, 10 atomic nuclei had been studied: ^{48}Ca, ^{76}Ge, ^{82}Se, ^{96}Zr, ^{100}Mo, ^{116}Cd, ^{128}Te, ^{130}Te, ^{136}Xe, and ^{150}Nd.

matrix element calculations, which were uncertain. The GERDA collaboration could do the comparison in a model-independent way because they were using the same isotope. They summarized their results as follows and included as well as comparison with the results reported in Klapdor-Kleingrothaus et al. (2004a). And here they pointedly noted that "[b]ecause of inconsistencies in [Klapdor-Kleingrothaus & Krivosheina, 2006] pointed out recently (Schwingenheuer, 2013, p. 273), the present comparison is restricted to the result of Klapdor-Kleingrothaus et al. (2004a)" (122503-2).

> The GERDA data show no indication of a peak at $0\nu\beta\beta$, i.e., the claim for the observation of $0\nu\beta\beta$ decay in ^{76}Ge is not supported. Taking $T^{0\nu}_{1/2}$ from [Klapdor-Kleingrothaus et al., 2004a] at its face value, 5.9 ± 1.4 decays are expected … in $\Delta E = \pm 2\sigma_E$ and 2.0 ± 0.3 background events after the PSD [pulse shape discrimination] cuts, as shown in Fig. 5.3. This can be compared with three events detected, none of them within $Q_{\beta\beta} \pm \sigma_E$. The model ($H_1$), which includes the $0\nu\beta\beta$ signal calculated above, gives in fact a worse fit to the data than the background-only model (H_0) : the Bayes factor, namely the ratio of the probabilities of the two models, is $P(H_1)/P(H_0) = 0.024$. Assuming the model H_1, the probability to obtain $N^{0\nu}$ = 0 as the best fit from the profile likelihood analysis is $P(N^{0\nu} = 0|H_1) = 0.01$ (Agostini et al., 2013, p. 122503-4).

The GERDA group then combined these results with those of KamLAND and EXO-200, the latter needed nuclear matrix element calculations, and found a Bayes factor for $P(H_1)/P(H_0) = 0.0022$. They concluded, "The long-standing claim for a $0\nu\beta\beta$ signal in ^{76}Ge is strongly disfavored" (p. 122503-5).

A 2015 result on ^{130}Te was presented by the CUORE-0 collaboration, which used a TeO bolometer array (Alfonso et al., 2015).[20] They found that the observed peak, attributed to ^{60}Co double-gamma events reconstructed at a somewhat higher energy than expected. The double-escape peak of ^{208}Tl did, however, reconstruct at the expected value. "Since e^+e^- pairs and $0\nu\beta\beta$ decays share similar event topologies we assume the latter would reconstruct according to the calibrated energy scale" (4). They unblinded their data, removing the artificial peak. They found no evidence for $0\nu\beta\beta$ decays (Fig. 5.4). Note the peak due to ^{60}Co double-gamma events at about 2507 keV and the absence of events at approximately 2527 keV, the energy expected for neutrinoless double beta decay of ^{130}Te. The χ^2 for the data in Fig. 5.4 was 43.9 for 46 degrees of freedom, a good fit.[21] The group set a lifetime limit of $T^{0\nu}_{1/2} \geq 2.7 \times 10^{24}$ y. When they combined this result with that of CUORICINO they set a limit of $T^{0\nu}_{1/2} \geq 4.0 \times 10^{24}$ y. "In summary, CUORE-0 finds no evidence for $0\nu\beta\beta$ decay of ^{130}Te and, when combined with Cuoricino, achieves the most stringent limit to date on this process" (6). Their upper limit on the neutrino mass was < 270–760 meV.

Negative evidence continued to accumulate. At the XXVIII International Conference on Neutrino Physics and Astrophysics, the results of five second generation experiments

---

[20] This was an improved version of the CUORICINO experiment discussed earlier.

[21] A χ^2 analysis yields a probability of 56 percent that the result is not a statistical fluctuation.

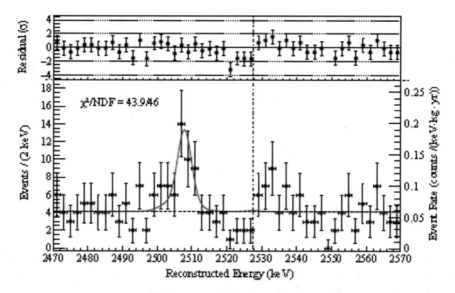

Fig. 5.4 Bottom: the best-fit model from the UEML fit (solid line) overlaid on the spectrum of $0\nu\beta\beta$ decay candidates in CUORE-0 (data points); the data are shown with Gaussian error bars. The peak at ~2507 keV is attributed to ^{60}Co; the dotted line shows the continuum background component of the best-fit model. Top: the normalized residuals of the best-fit model and the binned data. The vertical dot-dashed line indicates the position of $Q_{\beta\beta}$. *Source* Alfonso et al. (2015)

were reported (Table 5.1). None of these experiments reported evidence for neutrinoless double beta decay and they set more stringent limits on its presence. While these new results were still not sufficient to completely rule out the Heidelberg-Moscow results, they make that result extremely unlikely. The KamLAND-Zen group, for example stated that they had observed no significant $0\nu\beta\beta$ signal. Both GERDA and MAJORANA DEMON-STRATOR found one event in the appropriate energy region. These were consistent with background calculations and set more stringent limits on the decay. Tellingly none of these second generation experiments made any reference to the work of Klapdor-Kleingrothaus and his collaborators. Nor was there any mention made of that work in the presentations made at the XXVIII International Conference on Neutrino Physics and Astrophysics.[22]

Evidently the consensus was that delving any further into the data analysis offered by Klapdor-Kleingrothaus et al. was not pursuit worthy. This was to be expected given the consistent and mounting cascade of the negative experimental results. And while these negative results did not conclusively rule out the Heidelberg-Moscow positive result, they did offer persuasive evidence that it would not be a good use of time and resources

---

[22] Franklin's colleague, Alysia Marino reports that only one question was raised about the Heidelberg-Moscow results. The speaker, from the GERDA collaboration, did not have a slide available to answer the question, but remarked that they had excluded it with a high degree of confidence.

Table 5.1 Results presented
at the XXVIII Conference on
Neutrino Physics and
Astrophysics

Experiment	Isotope	$T^{0\nu}_{1/2}$ (10^{25} yr)	$m_{\beta\beta}$ (eV)
GERDA	^{76}Ge	> 8.0	< 0.12–0.26
MAJORANA	^{76}Ge	> 2.7	< 0.20–0.43
KamLAND-Zen	^{136}Xe	> 10.7	< 0.06–0.16
EXO	^{136}Xe	> 1.8	< 0.15–0.40
CUORE	^{130}Te	> 1.3	< 0.11–0.52
HM	^{76}Ge	$(2.23^{+0.44}_{-0.31})$	0.32 ± 0.03

As of June 2018

to delve into the statistical infirmities that may have affected the analysis offered by
Klapdor-Kleingrothaus et al.

Moreover, it was not just that no evidence had been obtained that the Majorana neutrino had been found, but importantly that the half-life and mass determinations became increasingly more certain and longer in duration. Which served to support the appraisal that these new experiments were in fact on the right track with regard to experimental refinement and development. On this point Table 5.1 shows the results of the five second generation experiments discussed above along with that of Klapdor-Kleingrothaus for the lifetimes for neutrinoless double beta decay along with the range of Majorana neutrino masses. (And here it should be kept in mind that the neutrino mass values depend on the nuclear matrix element.) As one can see GERDA and KamLAND-Zen set limits on the lifetime much longer that the HM value. The limits on the neutrino mass are also lower than the HM value although only the KamLAND-Zen result excludes the HM value.

The current situation with respect to the search for $0\nu\beta\beta$ has been concisely summarized in the *Annual Review of Nuclear and Particle Science* as follows.

> The experimental search for this decay is extremely challenging, and all previous attempts have returned empty-handed with the best current half-life limits of >10^{26} y. The experimental difficulties are matched by the theoretical ones; in particular, understanding the nuclear physics aspects of the decay has been a persistent challenge (Dolinski et al., 2019, p. 221).

Given this combination of difficulty and failure to detect, the question arises whether the search for $0\nu\beta\beta$ should be continued or whether it is time to shift resources, if only in part, to other ventures. What, in other words, is the shelf life of such pursuitworthiness? Before continuing we need to emphasize that what's involved here is the pursuit of an *experimental program* with the aim of detecting $0\nu\beta\beta$ and not just a single, narrowly defined experiment and where the boundaries of that program have been expanding to incorporate the raft of ongoing refinement and improvement.

Judging from the rather large number of such experiments under way or soon to be, the scientific consensus is that the search is still very much pursuitworthy.[23] There were, we think, three considerations that were taken to justify such continued pursuit. First and foremost, there was the great reward, that $0\nu\beta\beta$ once discovered could be leveraged to reveal what lies beyond the SM, and in particular the underlying symmetries that govern leptons. Add to this the considerable progress made regarding experimental technique and the corresponding promise that an experimental resolution of the existence of $0\nu\beta\beta$ might be at hand. Finally, there was the practical reality that such experimentation is "the only realistic direct probe for lepton-number violation" (Dolinski et al., 2019, p. 246). Continued pursuit thus was clearly in order.

References

Agostini, M., Allardt, M., Andreotti, E., Bakalyarov, A. M., Balata, M., Barabanov, I., Heider, M. B., Barros, N., Baudis, L., Bauer, C., & Becerici-Schmidt, N. (2013). Results on double-β decay of ^{76}Ge from phase 1 of the GERDA experiment. *Physical Review Letters, 111*, 122503-122501–122503-122506.

Alfonso, K., Artusa, D. R., Avignone III, F. T., Azzolini, O., Balata, M., Banks, T. I., Bari, G., Beeman, J. W., Bellini, F., Bersani, A., Biassoni, M., Brofferio, C., Bucci, C., Caminata, A., Canonica, L., Cao, X. G., Capelli, S., Cappelli, L., Carbone, L., ... Zucchelli, S. (2015). Search for neutrinoless double-beta decay of ^{130}Te with CUORE-0. *Physical Review Letters, 115*, 102502-102501–102502-102507.

Arnaboldi, C., Artusa, D. R., Avignone III, F. T., Balata, M., Bandac, I., Barucci, M., Beeman, J. W., Brofferio, C., Bucci, C., Capelli, S., & Carbone, L. (2005). New limit on the neutrinoless $\beta\beta$ decay of ^{130}Te. *Physical Review Letters, 95*, 142501-142501–142501-142504.

Avignone, F. T. (2008). Double beta decay, Majorana neutrinos, and neutrino mass. *Reviews of Modern Physics, 80*, 481–568.

Bakalyarov, A. M., Balysh, A. Y., Belyaev, S. T., Lebedev, V. I., & Zhukov, S. (2005). Results of the experiment on investigation of ^{76}Ge double beta decay. *Pisma Fiz. Elem. Chast. Atom., 2*, 21–28.

Barabash, A., Boehm, F., Brodzinski, R. L., Collar, J. I., Doe, P. J., Ejiri, H., Elliott, S. R., Fiorini, E., & Gaitskell, R. J. (2002a). Comment on evidence for neutrinoless double beta decay. *Modern Physics Letters A, 17*, 1475–1478.

Barabash, A. S. (2011). Experiment double beta decay: Historical review of 75 years of research. *Physics of Atomic Nuclei 74*, 603–613.

Baudis, L., Günther, M., Hellmig, J., Heusser, G., Hirsch, M., Klapdor-Kleingrothaus, H. V., Paes, H., Ramachers, Y., Strecker, H., Völlinger, M., & Bakalyarov, A. (1997). The Heidelberg-Moscow experiment: Improved sensitivity for ^{76}Ge neutrinoless double beta decay. *Physics Letters B, 407*, 219–224.

Dolinski, M. J., Poon, A. W. P., & Rodejohann, W. (2019). Neutrinoless double-beta decay: Status and prospects. *Annual Review of Nuclear and Particle Science, 69*, 219–251.

Elliott, S. R., Hahn, A. A., & Moe, M. K. (1987). Direct evidence for double beta-decay in ^{82}Se. *Physical Review Letters, 59*, 2020–2023.

Franklin, A. (2001). *Are there really neutrinos? an evidential history*. Cambridge, MA, Perseus Books.

---

[23] For a listing of fourteen such experiments and the associated technological advances see Dolinski et al. (2019, pp. 237–245).

Franklin, A. (2013). *Shifting standards: Experiments in particle physics in the twentieth century.* Pittsburgh, University of Pittsburgh Press.

Franklin, A. (2018). *Is it the same result? replication in physics.* San Rafael, CA, Morgan and Claypool.

Franklin, A., & Howson, C. (1984). Why do scientists prefer to vary their experiments? *Studies in History and Philosophy of Science, 15,* 51–62.

Franklin, A., & Laymon, R. (2021). *Once can be enough: Decisive experiments, no replication required.* Springer.

Furry, W. H. (1939). On transition probabilities in double beta-disintegration. *Physical Review, 56,* 1184–1193.

Giuliani, A., & Poves, A. (2012). Neutrinoless double-beta decay. *Advances in High Energy Physics, 2012,* 857016-857011–857016-857038.

Goeppert-Mayer, M. (1935). Double-beta disintegration. *Physical Review, 48,* 512–516.

Inghram, M. G., & Reynolds, J. H. (1950). Double beta-decay of Te^{130}. *Physical Review, 78,* 822–833.

Klapdor-Kleingrothaus, H. V., Dietz, A., Baudis, L., Heusser, G., Krivosheina, I. V., Majorovits, B., Paes, H., Strecker, H., Alexeev, V., Balysh, A., & Bakalyarov, A. (2001a). Latest results from the Heidelberg-Moscow double beta decay experiment. *European Physical Journal A, 12,* 147–154.

Klapdor-Kleingrothaus, H. V., Dietz, A., Harney, H. L., & Krivosheina, I. V. (2001b). Evidence for neutrinoless double beta decay. *Modern Physics Letters A, 16,* 2409–2420.

Klapdor-Kleingrothaus, H. V. (2002). Reply to a comment of article evidence for neutrinoless double beta decay. *arXiv* hep-ph: 0205228.

Klapdor-Kleingrothaus, H. V., Krivosheina, I. V., Dietz, A., & Chkvorets, O. (2004a). Search for neutrinoless double beta decay with enriched ^{76}Ge in Gran Sasso 1990–2003. *Physics Letters B, 586,* 198–212.

Klapdor-Kleingrothaus, H. V., Dietz, A., Krivosheina, I. V., & Chkvorets, O. (2004b). Data acquisition and analysis of the ^{76}Ge double beta experiment in Gran Sasso 1990–2003. *Nuclear Instruments and Methods in Physical Research A, 522,* 371–406.

Klapdor-Kleingrothaus, H. V., & Krivosheina, I. V. (2006). The evidence for the observation of 0νββ: The identification of 0νββ events from the full spectra. *Modern Physics Letters A, 20,* 1547–1566.

Majorana, E. (1937). Symmetrical theory of the electron and the positron. *Il Nuovo Cimento, 5,* 171–184.

Majorana, E. (2006). A symmetric theory of electrons and positrons. With translation and commentary by L. Maiani. In G. F. Bassani (Ed.), *Ettore scientific papers* (pp. 201–233). Springer.

Moe, M. (2014). The first direct observation of double-beta decay. *Annual Review of Nuclear and Particle Science, 64,* 247–267.

Pontecorvo, B. (1982). The infancy and youth of neutrino physics: Some recollections. *Journal de Physique, 43*(Suppl. C8), 221–236.

Primakoff, H. P. R. (1951). Angular correlation of electrons in double beta-decay. *Physical Review, 85,* 888–890.

Racah, G. (1937). Sulla simmetria tra particelle e antiparticelle. *Il Nuovo Cimento, 14,* 322–328.

Racah, G. (1982). On the symmetry between particles and antiparticles, translation by K. Nakagawa. *Soryushiron Kenkyu, 65,* 34–41.

Saakyan, R. (2013). Two-neutrino double-beta decay. *Annual Review of Nuclear and Particle Science, 63,* 503–529.

Schwingenheuer, B. (2013). Status and prospects of searches for neutrinoless double beta decay. *Annalen Der Physik, 525,* 269–280.

Zdesenko, Y. G., Danevich, F. A., & Tretyak, V. I. (2002). Has neutrinoless double β beta decay of ^{76}Ge been really observed? *Physics Letters B, 546,* 206–215.

Supersymmetry and the Expansion of the Standard Model

6

6.1 Supersymmetry and Its Application to the Standard Model

The Standard Model (SM), the currently accepted theory of elementary particle physics is one of the most successful theories ever. There is no currently known experimental result that conflicts with its predictions.[1] And it is complete insofar as the discovery of the Higgs boson in 2012 filled in the last remaining entity slot. Nevertheless, there has been considerable effort expended to search for physics beyond the SM. In particular, there is a problem, or rather a puzzle, concerning the observed mass of the Higgs boson. It's surprisingly small where its value 125 GeV, is far less than the mass scales one might expect, the Planck mass, 1.22×10^{19} GeV, or the Grand Unified Theory (GUT) scale, 1×10^{16} GeV. One proposed explanation involves the incorporation of the SM into the algebraic context known as Supersymmetry (SUSY).[2]

By way of background, the expression "Supersymmetry" is somewhat ambiguous as to its reference. Historically, it was a purely mathematical algebraic structure that looked as if it could be applied to a range of problems, primarily in physics, that would enable solutions to existing problems to be derived more readily or in some cases when no other methods were available. The basic idea was to identify a class of transformations (aka symmetries) such that if those transformations were satisfied by two different classes of entities, then results would follow as to more fundamental relationships between those

---

[1] As discussed in Chap. 7, there is now one tantalizing result that stands in conflict with the predictions of the SM.

[2] The Higgs boson plays a central role in the mechanism that generates the masses of the fundamental particles that make up the constituency of the SM, where that mass is a manifestation of potential energy transferred to these fundamental particles when they interact with the Higgs field. Thus, solving the puzzle of its smaller than anticipated mass is almost automatically pursuit worthy.

© The Author(s), under exclusive license to Springer Nature Switzerland AG 2022
R. Laymon and A. Franklin, *Case Studies in Experimental Physics*,
Synthesis Lectures on Engineering, Science, and Technology,
https://doi.org/10.1007/978-3-031-12608-6_6

entities.[3] The mathematics involved, however, is rather complex and we can do no more here than give the reader a rough idea as to what is involved. But there's a good analogy to be made with a theorem that we earlier reviewed in our discussion of the Wu experiment. There the transformation (or symmetry) was defined as a mirror image in the sense that the x, y, z location values of a system of interest were respectively transformed into $-x$, $-y$ and $-z$. The theorem held that if the image of a dynamical system (such as beta decay) was equivalent (in a certain defined sense) to its mirror image, then parity was conserved for that system. Otherwise, it was not conserved. As can be readily appreciated, this is a powerful result since it gives a relatively simple test for parity conservation.

The application of the supersymmetry transformation to the SM is similar in concept, where the classes of entities involved are (1) those of the SM, and (2) the elements of the SM as transformed by the symmetry requirements of SUSY. How this works is as follows:

> Elementary particles are divided into two classes; fermions, which have half-integral spin (1/2, 3/2, etc.), and bosons which have integral spin (0, 1, 2, etc.). Supersymmetry is the idea that each Standard Model particle has a supersymmetric partner, a sparticle, which is in the other class. Thus, we have the electron with spin ½ and the selectron with spin zero. Similarly, the photon, spin 1, has a supersymmetric partner, the photino with spin ½, and a spin ½ quark has a spin 0 squark (Sirunyan et al., 2017, p. 032003-1).

In sum, and with reference to the first iteration of SUSY as applied to the SM:

> In the minimal supersymmetric extension to the Standard Model, the so called MSSM, a supersymmetry transformation relates every chiral fermion and gauge boson in the SM to a supersymmetric partner with half a unit of spin difference, but otherwise with the same properties (such as mass) and quantum numbers (Buchmuller & de Jong, 2020, p. 1).

The dramatic and surprising payoff here is that upper limits on the mass of the Higgs Boson follow as a consequence.[4] This because:

> The extra particles predicted by supersymmetry would cancel out the contributions to the Higgs mass from their Standard-Model partners, making a light Higgs boson possible. The new particles would interact through the same forces as Standard-Model particles, but they would have different masses (CERN, 2022).[5]

---

[3] For a review of this history see Dimopoulos and Georgi (1981), Fayet (2001), and Rodriguez (2010).

[4] In addition to the success of SUSY with regard to the Higgs Boson, "it is intriguing that a weakly interacting, (meta)stable supersymmetric particle might make up some or all of the dark matter in the universe. In addition, SUSY predicts that gauge couplings, as measured experimentally at the electroweak scale, unify at an energy scale … near the Planck scale" (Buchmuller & de Jong 2020, p. 1).

[5] For the technical details regarding the underlying mechanism known as soft-electroweak symmetry breaking see Kane et al. (1993) and Espinosa and Quirós (1993).

For reasons that will become evident later, it's important to know that this application of SUSY to the SM with its dramatic payoff requires the specification of certain otherwise free parameters that play a central role in the derived limits on the Higgs boson. Thus,

> a judicious choice of the parameters in the theory is required to maintain the masslessness of the Higgs doublets while giving large mass to their colored SU(5) partners. ... Further, we note that any model of this kind requires a similar *fine tuning*. ... [Thus] the masslessness is not automatic or trivial (Dimopoulos & Georgi, 1981, pp. 151, 160, emphasis added).

It is also worth noting that there have been other examples in physics where an expansion of ontology has been achieved by taking advantage of available mathematical opportunities. The discovery of the positron and the continuing search for the magnetic monopole provide interesting similarities as compared with the application of SUSY to the SM. For both the positron and the magnetic dipole there was a mathematical theory that had unused variables where the instantiations for those variables suggested the existence of positrons and magnetic monopoles.

In the case of the positron, the Dirac equation had four solutions: Two of positive energy which accounted for the properties of electrons with opposite spin, and two of negative energy. The negative energy solutions were eventually found to correspond to Anderson's newly discovered positron.[6] It must be emphasized, however, that there was no necessity that positrons had to exist. If they didn't, then Dirac's negative solutions would have been no more than mathematical curiosities.

The possible existence of the magnetic monopole turns on the fact that Maxwell's equations have a mathematically symmetric set of twins, a sort of mirror image, where the electric and magnetic fields reverse roles whereby a moving magnetic monopole (the mirror image) is now accompanied by an electric field. Consequently, a moving monopole, if it existed, would create an electric field just as a moving charge creates a magnetic field. The fact that Maxwell's equations had such formally equivalent twins was already known in the nineteenth century, though the possibility that magnetic monopoles might have an independent existence was not taken seriously. Dirac thought otherwise. But, in this case, the search for their existence was not successful.[7]

The application of the mathematics of SUSY to the SM is similar insofar as the algebraic structures involved provided an opportunity for the insertion of entities not known to exist. But in the case of the SUSY and the SM, the proposed insertion is considerably more expansive since it involves the postulation of a mirror image (defined by the transformational symmetries involved) of the entire entity cohort of the SM.

Still, while the prediction of an upper bound on the mass of the Higgs boson was unexpected, that did not guarantee that the postulated partner entities of the SM actually

---

[6] For details Franklin and Laymon (2021, pp. 31–57).

[7] For details see Franklin and Laymon (2021, pp. 159–170). And for a recent example of a continuing search see Barrie et al. (2021).

exist. The resultant limitation on the mass of the Higgs Boson could after all have just been "accidental," just baked in, as it were, in the transformational properties assumed. There are two strategies to answer the question of whether the success here was merely accidental. Importantly, that these strategies are both currently being actively pursued is a testament to their *pursuit worthiness.*

The first is the obvious one of experimentally searching for the postulated partner entities. The second is less direct and involves the construction of supersymmetry formalisms that will produce more tightly defined constraints on the Higgs boson and other associated particles and processes, where those new constructions will in some sense be superior to the MSMM.[8]

Since our interests are primarily with the concepts of acceptance and pursuit as they apply to experimentation, we will focus, in what follows, on the first of these strategies, the search for the inhabitants of the supersymmetry mirror image of the SM.

6.2   Experimental Searches for the Supersymmetric Partners

The experimental search for sparticles is significantly complicated by the fact that MSSM by itself makes no predictions. It must be supplemented as follows:

> Since the mechanism by which SUSY is broken is unknown, a general approach to SUSY via the most general soft SUSY breaking Lagrangian *adds a significant number of new free parameters.* For the minimal supersymmetric standard model, MSSM, i.e., the model with the minimal particle content, these comprise 105 new real degrees of freedom. *A phenomenological analysis of SUSY searches leaving all these parameters free is not feasible.* For the practical interpretation of SUSY searches at colliders several approaches are taken to reduce the number of free parameters (Buchmuller & de Jong, 2020, p. 3, emphasis added).

Roughly speaking, once supplemented as described above, MSSM splinters into a gigantic array of possibilities, far too many to be experimentally searched. So, the problem is how to simplify while minimizing the inherent loss of reliability and precision due to such simplification. The need for supplementation to achieve an effective connection with the physical world is, of course, a feature of virtually every global theory. What distinguishes the case of MSSM is the extreme complexity involved and the uncertain cost of simplification.

The earliest approach to achieve effective simplification was "to assume a SUSY breaking mechanism and lower the number of free parameters through the assumption of additional constraints. … The most popular model was the constrained MSSM (CMSSM), which in the literature is also referred to as minimal supergravity, or MSUGRA." (Buchmuller & de Jong, 2020, p. 3). Because the constraints assumed were deemed too restrictive, this approach was superseded by one which employed,

---

[8] For a brief review of these developments see Allanach and Haber (2019, p. 25).

a broader and more comprehensive subset of the MSSM [which] is derived from the MSSM, using experimental data to eliminate parameters that are free in principle but have already been highly constrained by measurements of e.g., flavor mixing and CP-violation. This effective approach reduces the number of free parameters in the MSSM to typically 19 or even less, making it a practical compromise between the full MSSM and highly constrained models such as the CMSSM (p. 5).

There is currently yet another approach which relies on "simplified models" for the production of supersymmetric particles where,

[s]uch models assume a limited set of SUSY particle production and decay modes and leave open the possibility to vary masses and other parameters freely. Therefore, *simplified models enable comprehensive studies of individual SUSY topologies*, and are useful for optimization of the experimental searches over a wide parameter space without limitations on fundamental kinematic properties such as masses, production cross sections, and decay modes (p. 5, emphasis added).

Given the effectiveness and usefulness of this approach, "ATLAS and CMS have adopted such simplified models as the primary framework to provide interpretations of their searches" where "in practice, simplified model limits are often used as an *approximation* of the constraints that can be placed on sparticle masses in more complex SUSY spectra" (p. 5, emphasis added). But what determines whether such "approximations" are good enough for their intended purpose? And as expressly noted by Buchmuller and de Jong:

[D]epending on the assumed SUSY spectrum, the sparticle of interest, and the considered simplified model limit, this approximation can lead to a significant mistake, typically an overestimation, in the assumed constraint on the sparticle mass .... Only on a case-by-case basis can it be determined whether the limit of a given simplified model represents a good approximation of the true underlying constraint that can be applied on a sparticle mass in a complex SUSY spectrum (pp. 5–6).

In short, there is no easy answer to the question of whether such purported approximations are good enough for their intended purpose. To better appreciate what's involved in such determinations of the effectiveness of an approximation when SUSY is involved, we'll shift our attention to a particular example of a CMS search that relied on such a simplified model, where the individual SUSY topology investigated encompassed "squarks and gluinos produced in proton-proton ($p\,p$) collisions at $\sqrt{s} = 13$ TeV" (Sirunyan et al., 2017, 032003-1). But even though based on a simplified model, the experimental process was hardly simple.

6.3   The Search for Supersymmetry in Multijet Events with Missing Transverse Momentum in Proton-Proton Collisions at 13 TeV

The general scheme of these searches is to look for events that look like those expected from the production of sparticles.[9] There are several difficulties involved in these searches. Perhaps the most important is that ordinary Standard Model processes also produce events that mimic those expected from supersymmetry. The technique is to calculate the number of events that are produced by Standard Model processes and compare them to the number of observed events. This calculation will have not only a statistical uncertainty, but also systematic uncertainties. If the number of observed events is sufficiently larger than the number predicted by the Standard Model (usually expressed as a number of sigmas and requiring a 5σ effect), then one might claim the discovery of supersymmetric particles.

The experimenters began by noting the importance of the constraints imposed by Supersymmetry in possibly solving the fine-tuning and dark matter problems. "The standard model (SM) of particle physics describes many aspects of weak, electromagnetic, and strong interactions. However, it requires fine-tuning to explain the observed value of the Higgs boson mass, and it does not provide an explanation for dark matter. Supersymmetry (SUSY), a widely studied extension of the SM, potentially solves these problems through the introduction of a new particle, called a superpartner, for each SM particle, with a spin that differs from that of its SM counterpart by a half unit. Additional Higgs bosons and their superpartners are also introduced (p. 032003-1)." They also provided a brief description of the CMS apparatus, the same detector used in the discovery of the Higgs boson. "The central feature of the CMS apparatus is a superconducting solenoid of 6 m internal diameter, which provides a magnetic field of 3.8 T. Within the field volume are a silicon pixel and strip tracker, a lead tungstate crystal electromagnetic calorimeter (ECAL), and a brass/scintillator hadron calorimeter (HCAL). Muons are measured in gas-ionization detector embedded in the steel flux-return yoke. Extensive forward calorimeters complement the coverage provided by barrel and endcap detectors" (Chatrchyan et al., 2012, p. 31).

Another general problem in searching for physics beyond the Standard Model is the fact that although the LHC produces approximately 1.6×10^9 events/s[10] the CMS experiment can record only about 1000 events/s. Thus, the event recording rate is a factor of more than a million lower than the event production rate. The experimenters must have a way of deciding which events are of interest and should be recorded. This was done

---

[9] There have been numerous unsuccessful searches for physics beyond the Standard Model. As of 2018 the CMS collaboration at CERN had conducted 210 such searches. In addition, there were a similar number performed by the ATLAS collaboration.

[10] This depends, of course, on the intensity of the proton beams.

using trigger[11] and data acquisition systems. "The CMS trigger and data acquisition systems ensure that potentially interesting events are recorded with high efficiency" (p. 31). The CMS trigger was designed to perform that data reduction using different sequential triggers. "The first trigger level of CMS, Level-1, is hardware implemented and reduces the data rate, by using specific low level analysis in custom trigger processors. All further levels are software filters which are executed on (partial) event data in a processor farm. This is the upper level of real-time data selection and is referred to as High-Level Trigger (HLT). Only data accepted by the HLT are recorded for offline physics analysis" (Adam et al., 2006, p. 608). The HLT software can be changed relatively easily and there are various monitoring systems in place to ensure that such changes do not substantially change the operation of the experiment. "For this analysis, signal event candidates were recorded by requiring H_T^{miss} at the trigger level to exceed a threshold that varied between 100 and 120 GeV depending on the LHC instantaneous luminosity. The efficiency of this trigger, which exceeds 98% following application of the event selection criteria described below, is measured in data and is taken into account in the analysis. Additional triggers, requiring the presence of charged leptons, photons, or minimum values of H_T, are used to select samples employed in the evaluation of backgrounds, as described below" (Sirunyan et al., 2017, p. 032003-3).

Most of the events seen are produced by ordinary Standard Model interactions so further selection criteria were applied after the data were recorded to enhance any signal caused by sparticles. In this experiment there were six such criteria. The search assumed various simple models for the production of Supersymmetric particles. These models included final states which included top-antitop quark pairs, b-quarks and W and Z bosons, along with a pair of the Lightest Supersymmetric Particles (χ_1^0) (Fig. 6.1). Because quarks are confined within particles, they cannot exist in free form. They therefore fragment into hadrons before they can be directly detected, forming jets. B-quarks have a measurable, finite lifetime and can be distinguished from other jets because they begin a finite distance away from the primary vertex. A jet is a narrow cone of hadrons (strongly interacting particles) and other particles produced by a combination of quarks and gluons, the constituents of elementary particles. The event will also have missing transverse energy and momentum because the χ_1^0 is not observed.

The collaboration then described more details of their analysis procedures. Events considered as signal candidates are required to satisfy the following criteria:

(1) $N_{jet} \geq 2$, where jets must appear within $|\eta| < 2.4$.[12]
(2) $H_T > 300$ GeV, where H_T is the scalar p_T sum of jets with $|\eta| < 2.4$.

---

[11] A trigger system includes counters and other detectors along with computer programs for making a fast decision on whether to record the data for each event.

[12] η is a measure of how close the jet is to the forward direction where the detector does not have coverage. $\eta = 0$ corresponds to an angle of $90°$ with respect to the beam. $\eta = \infty$ is along the beamline.

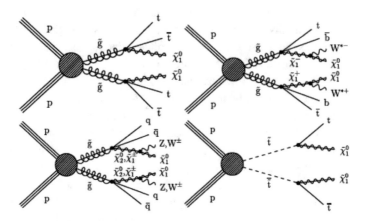

Fig. 6.1 Feynman diagrams for some of the simplified models considered in the search. *Source* Sirunyan et al. (2017)

(3) $H_T^{miss} > 300$ GeV, where H_T^{miss} is the magnitude of $\vec{H}_T^{miss}$, the negative of the vector p_T sum of jets with $|\eta| < 5$; an extended η range is used to calculate H_T^{miss} so that it better represents the total missing transverse momentum in an event.

(4) No identified, isolated electron or muon candidate with $p_T > 10$ GeV.

(5) No isolated track with $m_T < 100$ GeV and $p_T > 10$ GeV ($p_T > 5$ GeV if the track is identified as a PF electron or muon),[13] where m_T is the transverse mass formed from the $\vec{p}_T^{miss}$ and isolated-track p_T vector, where $\vec{p}_T^{miss}$ is the negative of the vector p_T sum of all PF objects.

(6) $\Delta\phi_{HmissTji} > 0.5$ for the two highest p_T jets j_1 and j_2, where $\Delta\phi_{HmissTji}$ is the azimuthal angle between $\vec{H}_T^{miss}$ and the p_T vector of jet j_i; if $N_{jet} \geq 3$, then, in addition $\Delta\phi_{HmissTj3} > 0.3$ for the third highest p_T jet j_3; if $N_{jet} \geq 4$, then, yet in addition, $\Delta\phi_{HmissTj4} > 0.3$ for the fourth highest p_T jet j_4; all considered jets must have $|\eta| < 2.4$ (p. 032003-3).

Once the data has been reduced by the "further selection" criteria a search was performed in a four-dimensional space with exclusive regions of N_{jet}, N_{b-jet}, H_T, and H_T^{miss}, where N_{jet} is the number of jets, N_{b-jet} is the number of tagged b-jets, H_T the scalar sum of the transverse momenta p_T of jets, and H_T^{miss} the magnitude of the vector p_T sum of jets. The search intervals in N_{jet} and N_{b-jet} were $N_{jet} = 2$, 3–4, 5–6, 7–8, ≥ 9 and $N_{b-jet} = 0$, 1, 2, ≥ 3. The intervals for H_T, and H_T^{miss} are shown in Fig. 6.2. Several intervals were discarded. "Events with both small H_T and large H_T^{miss} are not considered because such events are likely to arise from mismeasurement. For $N_{jet} \geq 7$, the kinematic intervals 1 and 4 are discarded because of the small number of events. The number of search regions is 174" (p. 032003-4). Regions C1, C2, and C3 are control regions which were used to

---

[13] PF is Particle Flow, an algorithm used to identify particles.

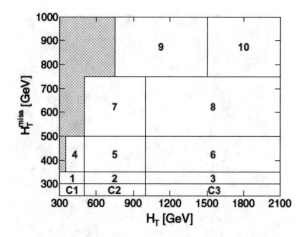

Fig. 6.2 Schematic illustration of the kinematic search intervals in the H_T^{miss} versus H_T plane. The intervals C1, C2, and C3 are the control regions used to estimate the QCD background. *Source* Sirunyan et al. (2017)

estimate the Standard Model background. A possible supersymmetry event is shown in Fig. 6.3. These regions have low H_T^{miss}, regions where no supersymmetric particles are expected. As discussed, earlier events of interest should contain jets as well as missing transverse momentum. The event shown in Fig. 6.3 has 12 jets, 3 b-quark jets and a missing energy of 671 GeV.

The estimation of background is crucial in this search. The experimenters used the data in the control regions as their primary data for estimating background. In addition, "Samples of simulated SM [Standard Model] events are used to validate the analysis procedures and for some secondary aspects of background estimation. The background processes considered included, top quark and W boson + jet events, which were further subdivided into lost-lepton background and hadronically decaying τ lepton background" (p. 032003-8). Other processes considered were the decay of the Z boson to a neutrino-antineutrino pair and QCD events.[14] "Uncertainties in the event yields resulting from the calibration of jet energy scale are estimated by shifting the jet momenta *in the simulation* up and down by one standard deviation of the jet energy corrections" (p. 032003-8). This is similar to, but not identical to, the technique discussed earlier in which possible confounding experimental effects are magnified to see if they had any significant effect on the final result. In this case the technique was applied to the simulation, not to the physical experimental apparatus. The experimenters found good agreement between the computer simulation and the data. "The lost-lepton background in the 174 search regions of the analysis as determined directly from $t^{bar}t$ [antitop-top], single top quark, W +

---

[14] The estimation of background is quite technical and will not be discussed here.

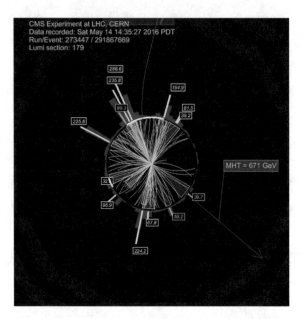

Fig. 6.3 A possible supersymmetry event. The event has 12 jets (yellow triangles) and 3 b-jets (blue triangles). The missing transverse energy is 671 GeV. *Courtesy* Kevin Stenson

jets, diboson, and rare-event simulation and as preduon control samples." (p. 0320003-6). Similar comparisons were made for other backgrouicted by applying the lost-lepton background determination procedure to simulated electron and mnds.

The experimental results are shown in Fig. . The data are shown by dots and the co6.4lored bins show the estimated background. The upper plot shows the number of observed events compared with the predictions for Standard Model processes. The lower plot shows (Number observed − Number Expected)/Number Expected for each of the search bins. This ratio should be zero if there is no significant supersymmetry signal. "Signal region 126 exhibits a difference of 3.5 standard deviations with respect to the SM expectation. Signal regions 74, 114, and 151 exhibit differences between 2 and 3 standard deviations. (These regions are defined by values of H_T^{miss}, H_T, N_{jet}, and $N_{b\text{-}jet}$.) The differences for all other signal regions lie below 2 standard deviations. Thus, the evaluated SM background is found to be statistically compatible with the data *and we do not obtain evidence for supersymmetry*" (p. 032003-11, emphasis added). Numerical values for the number of events observed and predicted for bins 126, 74, 114, and 151 are shown in Fig. 6.5. Figure 6.6 shows the number of observed and predicted events for 10 of the remaining 170 bins, a representative sample of the results. No large deviations, either positive or negative, are seen. The reader will note that the 3.5 standard deviation effect observed in bin 126 might qualify for an "evidence for" statement. This would be correct if the experimenters had predicted an effect in bin 126. The probability of a 3.5-standard-deviation effect in a single bin is 0.000465. The probability, however,

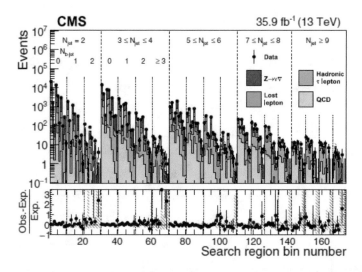

Fig. 6.4 The observed numbers of events and the prefit SM background predictions. The lower panel displays the fractional differences between the data and the SM predictions. Note that the data is both above and below the predictions. *Source* Sirunyan et al. (2017)

Bin	H_T^{miss} [GeV]	H_T [GeV]	N_{jet}	$N_{b\text{-}jet}$	Lost-e/μ	$\tau \to$ had	$Z \to \nu\bar{\nu}$	QCD	Total pred.	Obs.
74	350–500	350–500	5–6	0	71^{+7+11}_{-6-11}	47^{+3+16}_{-3-16}	242^{+9+20}_{-9-19}	$12.7^{+2.3+5.3}_{-2.3-4.8}$	372^{+13+29}_{-13-28}	464
114	350–500	>1000	7–8	0	$22.5^{+2.8+2.7}_{-2.7-2.7}$	$23.3^{+2.5+2.3}_{-2.4-2.3}$	$48.3^{+4.7+5.4}_{-4.3-4.8}$	$12.6^{+0.7+5.2}_{-0.7-4.8}$	$106.7^{+7.1+8.3}_{-6.7-7.7}$	75
126	>750	>1500	7–8	1	$0.00^{+0.47+0.00}_{-0.00-0.00}$	$0.14^{+0.47+0.02}_{-0.08-0.02}$	$0.70^{+0.55+0.22}_{-0.34-0.21}$	$0.03^{+0.01+0.01}_{-0.01-0.01}$	$0.9^{+1.1+0.2}_{-0.3-0.2}$	6
151	300–350	500–1000	≥9	1	$6.5^{+1.8+1.1}_{-1.7-1.1}$	$4.57^{+0.93+0.77}_{-0.81-0.77}$	$1.83^{+0.84+0.68}_{-0.60-0.74}$	$1.02^{+0.06+0.42}_{-0.06-0.40}$	$13.9^{+2.8+1.5}_{-2.6-1.6}$	25

Fig. 6.5 The four bins with the largest statistical excess of observed events compared with predicted events. *Source* Sirunyan et al. (2017)

of observing one 3.5 sigma effect anywhere in 174 bins is, in fact, 0.922.[15] We may explain this result quite simply. It is true that the probability of observing a 3.5 σ effect in a single bin is 0.000465. The probability of *not* observing such an effect in a single bin is $1 - 0.000465 = 0.999535$. The probability of not observing the effect in two bins is 0.999535×0.999535, and in n bins it is $(0.999535)^n$, assuming the bins are probabilistically independent. For 174 bins this gives $(0.999535)^{174} = 0.078$. Therefore, the probability of *observing* a 3.5σ effect in 174 bins is $1 - 0.078 = 0.922$. Thus, the collaboration was not surprised to observe such an effect. In this case the experimenters are arguing for a null result based on the statistical disconfirmation of a test hypothesis, the existence of supersymmetric particles.

---

[15] This is known as the "look elsewhere" effect.

Bin	H_T^{miss} [GeV]	H_T [GeV]	N_{jet}	$N_{b\text{-}jet}$	Lost-e/μ	$\tau \to$ had	$Z \to \nu\bar{\nu}$	QCD	Total pred.	Obs.
1	300–350	300–500	2	0	$4069^{+67+320}_{-67-320}$	$2744^{+37+510}_{-37-500}$	$13231^{+67+760}_{-66-740}$	$326^{+12+170}_{-12-120}$	$20370^{+120+980}_{-120-960}$	21626
11	300–350	300–500	2	1	370^{+21+31}_{-21-31}	288^{+11+63}_{-11-63}	1361^{+7+140}_{-7-140}	44^{+6+25}_{-6-17}	$2063^{+33+160}_{-33-160}$	1904
32	300–350	500–1000	3–4	0	$1125^{+25+120}_{-25-120}$	$909^{+18+100}_{-18-100}$	$2487^{+29+140}_{-28-140}$	119^{+8+51}_{-8-45}	$4640^{+52+220}_{-52-210}$	4799
41	300–350	300–500	3–4	1	746^{+25+55}_{-25-55}	627^{+15+48}_{-15-47}	1235^{+8+130}_{-8-120}	59^{+4+24}_{-4-22}	$2667^{+41+150}_{-41-150}$	2677
52	300–350	500–1000	3–4	2	$92.3^{+9.1+9.5}_{-9.0-9.5}$	$85.6^{+5.7+7.5}_{-5.7-7.4}$	$53.0^{+0.6+9.6}_{-0.6-9.6}$	$3.8^{+1.2+1.6}_{-1.2-1.4}$	235^{+15+16}_{-15-15}	227
71	300–350	300–500	5–6	0	217^{+11+22}_{-11-22}	166^{+6+27}_{-6-27}	489^{+12+42}_{-12-39}	49^{+5+21}_{-5-19}	922^{+21+58}_{-21-56}	1015
82	300–350	500–1000	5–6	1	290^{+11+25}_{-11-25}	302^{+8+25}_{-8-25}	218^{+4+31}_{-4-30}	41^{+4+17}_{-4-16}	851^{+20+50}_{-20-49}	781
86	350–500	>1000	5–6	1	$40.5^{+5.5+4.2}_{-5.4-4.2}$	$36.0^{+3.3+4.3}_{-3.3-4.2}$	$49.4^{+2.3+7.0}_{-2.2-6.7}$	$11.9^{+0.7+4.8}_{-0.7-4.5}$	138^{+9+10}_{-9-10}	138
111	300–350	500–1000	7–8	0	$48.0^{+3.9+5.4}_{-3.8-5.4}$	$60.8^{+3.4+6.0}_{-3.4-6.0}$	76^{+5+11}_{-5-10}	30^{+2+12}_{-2-11}	215^{+9+18}_{-9-17}	218
112	300–350	>1000	7–8	0	$21.2^{+2.9+2.3}_{-2.9-2.3}$	$20.3^{+2.2+2.8}_{-2.1-2.8}$	$23.9^{+3.3+2.8}_{-2.9-2.5}$	$20.5^{+0.5+8.5}_{-0.5-7.8}$	$85.9^{+6.1+9.6}_{-5.8-9.0}$	85

Fig. 6.6 Number of observed events compared with the number of predicted events for 10 bins. *Source* Sirunyan et al. (2017)

The collaboration was also able to use their null result to set lower limits, at the 95 percent confidence level, on the mass of the gluino and the mass of the squark. "Using the predicted cross-sections with next-to-leading order plus next-to-leading logarithm accuracy as a reference, 95% confidence lower limits on the gluino mass as large as 1800–1960 GeV were derived, depending on the scenario (Fig. 6.7). The corresponding limits on the mass of directly produced squarks range from 960 to 1390 GeV" (p. 032003-16–032003-17).

6.4   Pursuit Worthiness Versus Failure to Deliver on the Promise

Viewed individually—and putting aside that what was achieved was at best an "approximation" of uncertain accuracy—the experiment was a success insofar as it restricted the available real estate that the sparticles in question could occupy. Moreover, its results are consistent with the restrictions derived from other experiments dealing with different SUSY topologies. Given this consistency, Buchmuller and de Jong concluded in particular that the "[t]he absence of any observation of new [i.e., contradictory] phenomena at the first run of the LHC at $\sqrt{s} = 7/8$ TeV, and after the second run at $\sqrt{s} = 13$ TeV, place significant constraints on SUSY parameter space" (2020, pp. 22–23). So, progress has been made. And here we note that this progress, in the sense of the development of

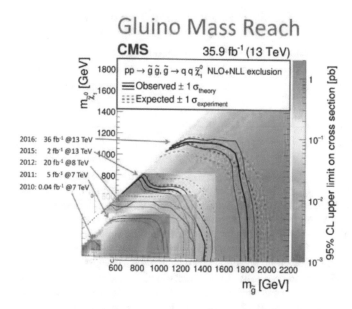

Fig. 6.7 The lower limits on the on the masses of the χ_1^0, the lowest mass supersymmetric particle, and the gluino, the supersymmetric counterpart of the gluon, set by the CMS collaboration from 2010 to 2016. *Courtesy* Keith Ulmer

consistent and more restrictive constraints on parameter space, is some indication as to the adequacy of the approximations achieved by the use of simplified models.[16]

On the other hand, none of these experiments detected any of the sought after sparticles, and so in that sense were not successful. And after carefully reviewing what is now an *extensive array of such negative results*, Allanach and Haber made the following somewhat gloomy assessment:

> At present, there is no direct evidence for weak-scale SUSY from the data analyzed by the LHC [CERN Large Hadron Collider] experiments. … In light of these *negative results*, one must confront the tension that exists between the theoretical expectations for the magnitude of the SUSY-breaking parameters and the non-observation of supersymmetric phenomena at colliders (Allanach & Haber, 2019, p. 19, emphasis added).

In other words, there's disconnect between the initially determined pursuit worthiness and the consequent failure to deliver on that promise. One way to confront the disconnect is, as we mentioned above, to continue the theoretical search for better models and versions

---

[16] There are, however, some instances which indicate "that care must be taken when interpreting results from the LHC searches and there are still several scenarios where sparticles below the 1 TeV scale are not excluded, even when considering the most recent results at $\sqrt{s} = 13$ TeV" (Buchmuller & de Jong, 2020, p. 23).

of SUSY. The other is to continue, with improved sensitivity, the search for the SUSY companion particles. See Buchmuller and de Jong (2020, pp. 23–24) for a review of both ongoing and planned experimental efforts, but where "the improvement in sensitivity will largely have to come from a larger data set, and evolution of trigger and analysis techniques, since there will be no significant energy increase at the LHC anymore."

References

Adam, W., Bergauer, T., Deldicque, C., Erö, J., Fruehwirth, R., Jeitler, M., Kastner, K., Kostner, S., Neumeister, N., Padrta, M., Porth, P., Rohringer, H., Sakulin, H., Strauss, J., Taurok, A., Walzel, G., Wulz, C. E., Lowette, S., van de Vyver, B., de Lentdecker, G., Vanlaer, P., Delaere, C., Lemaitre, V., Ninane, A., van der Aa, O., Damgov, J., Karimäki, V., Kinnunen, R., … Smith, W. H. (2006). The CMS high level trigger. *European Physics Journal C, 46*, 605–667.

Allanach, B. B., & Haber, H. E. (2019) Supersymmetry, Part I (Theory). Available at the Particle Data Group (PDG) website: https://pdg.lbl.gov/2019/reviews/rpp2019-rev-susy-1-theory.pdf

Barrie, N. D., Sugamoto, A., Talia, M., & Yamashita, K. (2021). Searching for monopoles via monopolium multiphoton decays. *Nuclear Physics B, 972*, 115564.

Buchmuller, O., & de Jong, P. (2020). Supersymmetry, Part II (Experiment). Available at the Particle Data Group (PDG) website: https://pdg.lbl.gov/2020/reviews/rpp2020-rev-susy-2-experiment.pdf

CERN. (2022). Supersymmetry predicts a partner particle for each particle in the Standard Model, to help explain why particles have mass. Available at: https://home.cern/science/physics/supersymmetry

Chatrchyan, S., Khachatryan, V., Sirunyan, A. M., Tumasyan, A., Adam, W., Aguilo, E., Bergauer, T., Dragicevic, M., Erö, J., Fabjan, C., & Friedl, M. (2012). Observation of a new boson at a mass of 125 GeV with the CMS experiment at the LHC. *Physics Letters B, 716*, 30–61.

Dimopoulos, S., & Georgi, H. (1981). Softly Broken Supersymmetry and SU(5). *Nuclear Physics, B, 193*, 150–161.

Espinosa, J. R., & Quirós, M. (1993). Upper bounds on the lightest Higgs boson mass in general supersymmetric Standard Model. *Physics Letters B, 302*, 51–58.

Fayet, P. (2001). About the origins of the supersymmetric Standard Model. *Nuclear Physics B: Proceedings Supplements, 101*, 81–98.

Franklin, A., & Laymon, R. (2021). *Once can be enough: Decisive experiments, no replication required*. Springer.

Kane, G. L., Kolda, C., & Wells, J. D. (1993). Calculable upper limit on the mass of the lightest Higgs boson in perturbatively valid supersymmetric theories with arbitrary Higgs sectors. *Physical Review Letters, 70*, 2686.

Rodriguez, M. C. (2010). History of supersymmetric extensions of the Standard Model. *International Journal of Modern Physics A, 25*, 1091–1121.

Sirunyan, A. M., Tumasyan, A., Adam, W., Ambrogi, F., Asilar, E., Bergauer, T., Brandstetter, J., Brondolin, E., Dragicevic, M., Erö, J., & Flechl, M. (2017). Search for supersymmetry in multijet events with missing transverse momentum in proton-proton collisions at 13 TeV. *Physical Review D, 96*, 032002-032001–032003-032038.

The Anomalous Magnetic Moment of the Muon

7

One of the great successes of quantum electrodynamics, a part of the Standard Model, has been the prediction of the g-factor of both the electron and the muon. In fact, the calculated and measured values of the electron g-factor agree to within a part in a trillion.[1] In 2001, however, a small, but tantalizing difference was found between the calculated and measured values of the g-factor of the muon.

In this episode we will discuss pursuit of this discrepancy, a pursuit which involved considerable effort and expense in moving the experimental apparatus more than 3000 miles. This was in addition to the difficulties involved in making a precision measurement. Considerable effort was also made in performing the theoretical calculations to be compared with the experimental result. In 2001 a group working at the Alternate Gradient Synchrotron at Brookhaven National Laboratory the (Muon (g − 2) Collaboration) reported their results based on data taken in 1999. They stated their purpose clearly.

> Precise measurement of the anomalous g value, $a_\mu = (g − 2)/2$, of the muon *provides a sensitive test of the standard model of particle physics and new information on speculative theories beyond it.* Compared to the electron, the muon g value is more sensitive to standard model extensions, typically by a factor of $(m_\mu/m_e)^2$ (Brown et al., 2001, p. 2227, emphasis added).

In this experiment positive, polarized muons (their spins are aligned) were injected into a storage ring, in which the muons were kept in orbit by a magnetic field <B>. The muon spin processes faster than its momentum rotates in the magnetic field by an angular frequency ω_a. The anomalous magnetic moment of the muon a_μ is given by

---

[1] A *g*-factor (also called *g* value or dimensionless magnetic moment) is a dimensionless quantity that characterizes the magnetic moment and angular momentum of an atom, a particle or the nucleus. It is essentially a proportionality constant that relates the different observed magnetic moments μ of a particle to their angular momentum quantum numbers and a unit of magnetic moment (to make it dimensionless), usually the Bohr magneton or nuclear magneton. See Aoyama et al. (2007).

© The Author(s), under exclusive license to Springer Nature Switzerland AG 2022   135
R. Laymon and A. Franklin, *Case Studies in Experimental Physics*,
Synthesis Lectures on Engineering, Science, and Technology,
https://doi.org/10.1007/978-3-031-12608-6_7

$$a_\mu = \omega_a / \left[(e/m_\mu c) <B> \right]$$

The angular frequency ω_a was determined by counting the number of positrons resulting from muon decay, $\mu^+ \rightarrow e^+ + v_e + v_\mu$. Parity violation in the decay[2] gives rise
to asymmetries in both the electron angular distribution and in the electron energy. The
number of positrons with energy greater than E is given by

$$N(t) = N_o e^{-t/(\gamma\tau)} \{ 1 + A(E) \sin[\omega_a t + \varphi_a(E)] \}$$

where $\gamma\tau$ is the time dilated lifetime of the muon, $A(E)$ is a function of the positron
energy and $\varphi_a(E)$ is a phase depending on the positron energy. Thus, by measuring the
distribution of decay positrons one can determine ω_a (Fig. 7.1). The other crucial parameter in determining a_μ is the magnetic field $<B>$. The magnetic field was measured by
seventeen nuclear magnetic resonance (NMR) probes mounted on a trolley which moved
on a fixed track inside the muon storage ring vacuum chamber. The measurements were
made approximately every three days, and interpolation for the period between measurements was provided by 150 fixed NMR probes distributed around the ring. Because of its
importance, the fitting of the magnetic field was delegated to four different and independent subgroups.[3] Each group used a different random offset in the value of the magnetic
field. One of the groups had difficulty in obtaining an internally consistent fit to the field
and in order to try to locate the problem the analysis was partially unblinded. Each group
was then given the same random offset, allowing a comparison of the fits. The problem was found and corrected. I note that the absolute value of the field, needed for the
calculation of a_μ, was still unknown to the experimenters (Brown et al., 2001).

Interestingly, the experiment–theory comparison was, at this time, in a sense, also
blind. After the experimental group had obtained their final value of a_μ, they asked
William Marciano, a theoretical physicist who worked on calculating its value, what was
the best theoretical value of a_μ. They did not inform him of their result until after he
had provided the value. Their experimental value was $a_\mu = 11{,}659{,}202$ (14)(6) $\times 10^{-10}$,
which was in good agreement with previously measured values, but with an experimental uncertainty only one third the size.[4] The best theoretical value, calculated from the
Standard Model, was $a_\mu = 11{,}659{,}159.6(6.7) \times 10^{-10}$. The difference $a_\mu(\text{exp}) - a_\mu(\text{SM})$
$= 43(16) \times 10^{-10}$, which might indicate a problem with the Standard Model, and was,
perhaps, an indication of the presence of supersymmetry, a currently favored theoretical
speculation. Figure 7.2 shows the results obtained by the Brookhaven group in 1997,
1998, and 1999, previous results from CERN, along with the theoretical prediction. One
can see the small discrepancy between theory and experiment.

---

[2] For details of parity nonconservation in the weak interactions see Chap. 3 of this book.

[3] The information about the blind analysis was presented by Gerry Bunce in a seminar at the
University of Colorado and in private conversation. It does not appear in the published paper.

[4] (14) and (6) were the statistical and systematic uncertainties in a_μ, respectively.

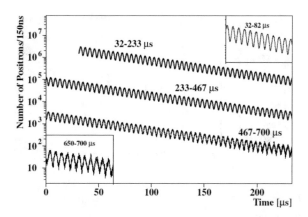

Fig. 7.1 Positron time spectrum overlaid with the fitted 10 parameter function (χ^2 per degree of freedom $= 3818/3799$). The total event sample of 0.95×10^9 with $E \geq 2.0$ GeV is shown. *Source* Brown et al. (2001)

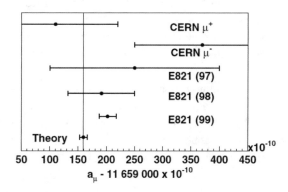

Fig. 7.2 Measurements of a_μ and the standard model prediction with their total errors. *Source* Brown et al. (2001)

In 2006 the Muon (g − 2) Collaboration published the "Final report of the E821 muon anomalous magnetic moment measurement at BNL" (Bennett et al., 2006).[5]

We present the final report from a series of precision measurements of the muon anomalous magnetic moment, $a_\mu = (g - 2)/2$. The details of the experimental method, apparatus, data taking, and analysis are summarized. Data obtained at Brookhaven National Laboratory, using nearly equal samples of positive and negative muons, were used to deduce a_μ (Expt) $= 11659208.0(5.4)\,(3.3) \times 10^{-10}$, where the statistical and systematic uncertainties are given, respectively. [The results are shown in Fig. 7.3]. The

---

[5] The group had published updates in 2002 and 2004.

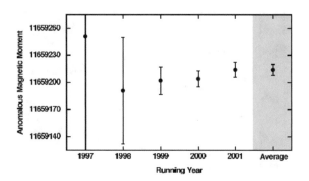

Fig. 7.3 Results for the E821 individual measurements of a_μ by running year, together with the final average. *Source* Bennett et al. (2001)

combined uncertainty of 0.54 ppm represents a 14-fold improvement compared to previous measurements at CERN. The standard model value for a_μ includes contributions from virtual QED, weak, and hadronic processes. While the QED processes account for most of the anomaly, the largest theoretical uncertainty, ≈ 0.55 ppm, is associated with first-order hadronic vacuum polarization. Present standard model evaluations, based on e^+e^- hadronic cross sections, lie 2.2–2.7 standard deviations below the experimental result (p. 072003-1).

The group noted not only improvements in the precision of the measurement, but also that the theoretical value had changed. They remarked, "Continued theoretical modelling of the hadronic light-by-light contribution can also be expected, and initial studies using the lattice have begun. We are confident that the precision on the standard model value will be improved, enabling a more sensitive comparison to experiment (072003-39)." They also remarked that their result had been limited by statistics. They suggested several experimental improvements that could be made and also stated that improvements in the accuracy and precision of the theoretical prediction was expected. "Thus, we may expect a significantly improved sensitivity for the anomaly test in the future. In the era of the LHC and direct searches for specific standard model extensions, precision measurements, such as that of the muon anomaly, represent a continually improving sum rule of known physics and provide independent insight into physics at high energies and short-distance scales" (p. 072300-39).

The two-standard-deviation discrepancy between theory and experiment raised the possibility of hints toward physics beyond the Standard Model and generated considerable pursuit in both theory and experiment. It would, however, take fourteen years before new results were available. In 2017 more than 100 theoretical physicists formed the Muon (g-2) Theoretical Initiative to provide more detailed calculations of the muon g-2. In 2020 their work resulted in a long, detailed, and comprehensive report, "The anomalous magnetic moment of the muon in the Standard Model" (Aoyama et al., 2020). This review was 166 pages long, contained 824 references, and more than 100 authors. The authors remarked

This more tantalizing discrepancy is not at the discovery threshold.[6] Accordingly, two major initiatives are aimed at resolving whether new physics is being revealed in the precision evaluation of the muon's magnetic moment. The first is to improve the experimental measurement of a^{exp}_μ by a factor of 4. The Fermilab Muon g − 2 collaboration is actively taking and analyzing data using proven, but modernized, techniques that largely adopt key features of magic-momenta storage ring efforts at CERN and BNL. [This will be discussed below].... The goal of the second effort is to improve the theoretical SM evaluation to a level commensurate with the experimental goals. To this end, a group was formed – the Muon g − 2 Theory Initiative – to holistically evaluate all aspects of the SM and to recommend a single value against which new experimental results should be compared. This White Paper (WP) is the first product of the Initiative, representing the work of many dozens of authors (Aoyama et al., 2020, pp. 5–6).

The result of this enormous effort was "$\Delta a_\mu = a^{exp}_\mu - a^{SM}_\mu = 279(76) \times 10^{-11}$, corresponding to a 3.7σ discrepancy (p. 148)."[7]

There had also been significant work on the experimental side. It was decided to move the experiment to Fermilab, whose accelerator had a higher proton beam intensity and would produce more muons. It was also decided to move the 14.2-m diameter magnetic storage ring from Brookhaven National Laboratory on Long Island to Fermilab, in Batavia, Illinois, rather than build a new ring at Fermilab. This involved a 3200-mile journey by specially constructed trucks and barge down the east coast of the U.S. and up the Mississippi River (Figs. 7.4 and 7.5).[8] Data taking began in 2017. "The experiment follows the BNL concept and uses the same 1.45 T superconducting storage ring (SR) magnet, but it benefits from substantial improvements. These include a 2.5 times improved magnetic field intrinsic uniformity, detailed beam storage simulations, and state of-the-art tracking, calorimetry, and field metrology for the measurement of the beam properties, precession frequency, and magnetic field (Abi et al., 2021, p. 141801-3)." Their final result was $a_\mu(FNAL) = 116592040(54) \times 10^{-11}$ (0.46 ppm). "Our result differs from the SM value by 3.3σ and agrees with the BNL E821 result. The combined experimental (Exp) average is $a_\mu(Exp) = 116592061(41) \times 10^{-11}$ (0.35 ppm). The difference, $a_\mu(Exp) - a_\mu(SM) = (251 \pm 59) \times 10^{-11}$, has a significance of 4.2σ. These results are displayed in Fig. 7.6 (p. 141801-7)." The difference was larger, but still not statistically significant enough to claim a discovery. This was, however, the first of five planned experimental runs. These would improve the statistical uncertainty by a factor of four.

---

[6] At this time a 5-sigma effect was needed to make a discovery claim.

[7] This was based on the 2004 result published by the Brookhaven group not the final 2006 result.

[8] This move took place in 2013 and took 35 days.

Fig. 7.4 The 50-foot-wide Muon g-2 electromagnet being driven north on I-355 between Lemont and Downers Grove, Illinois, shortly after midnight on Thursday, July 25, 2013. *Credit: Fermilab*

Fig. 7.5 The barge transporting the Muon g-2 electromagnet goes through locks on the Illinois River in Joliet, Illinois on Saturday, July 20, 2013. *Credit: Fermilab*

Interestingly, on the same day that the Fermilab experimental result was published, April 7, 2021, another theoretical calculation of g-2 appeared online (later published as Borsanyi et al., 2021). The group summarized the situation as follows

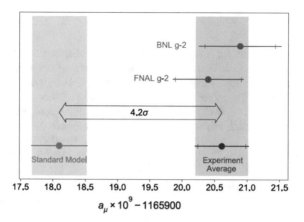

Fig. 7.6 From top to bottom: experimental values of a_μ from BNL E821, this measurement, and the combined average. The inner tick marks indicate the statistical contribution to the total uncertainties. The Muon $g-2$ Theory Initiative recommended value for the standard model is also shown. *Source* Abi et al. (2021)

The standard model of particle physics describes the vast majority of experiments and observations involving elementary particles. Any deviation from its predictions would be a sign of new, fundamental physics. One long-standing discrepancy concerns the anomalous magnetic moment of the muon, a measure of the magnetic field surrounding that particle. Standard-model predictions exhibit disagreement with measurements that is tightly scattered around 3.7 standard deviations. Today, theoretical and measurement errors are comparable; however, ongoing and planned experiments aim to reduce the measurement error by a factor of four. Theoretically, the dominant source of error is the leading-order hadronic vacuum polarization (LO-HVP) contribution. For the upcoming measurements, it is essential to evaluate the prediction for this contribution with independent methods and to reduce its uncertainties. The most precise, model-independent determinations so far rely on dispersive techniques, combined with measurements of the cross-section of electron–positron annihilation into hadrons. To eliminate our reliance on these experiments, here we use ab initio quantum chromodynamics (QCD) and quantum electrodynamics simulations to compute the LO-HVP contribution. We reach sufficient precision to discriminate between the measurement of the anomalous magnetic moment of the muon and the predictions of dispersive methods. *Our result favours the experimentally measured value over those obtained using the dispersion relation.* Moreover, the methods used and developed in this work will enable further increased precision as more powerful computers become available (Borsanyi et al., 2021, p. 51).

Thus, although the precision of the experimental value of the muon g-2 factor had improved, there was not only a discrepancy between experiment and theory, but there was also a difference between two different theoretical calculations. Pursuit continues on both the experimental and theoretical fronts.

References

Abi, B., Albahri, T., Al-Kilani, S., Allspach, D., Alonzi, L. P., Anastasi, A., Anisenkov, A., Azfar, F., Badgley, K., Baeßler, S., & Bailey, I. (2021). Measurement of the positive muon anomalous magnetic moment to 0.46 ppm. *Physical Review Letters, 126*, 141801-141801–141801-141811.

Aoyama, T., Asmussen, N., Benayoun, M., Bijnens, J., Blum, T., Bruno, M., Caprini, I., Calame, C. C., Cè, M., Colangelo, G., & Curciarello, F. (2020). The anomalous magnetic moment of the muon in the Standard Model. *Physics Reports (Section C of Physics Letters), 887*, 1–166.

Aoyama, T., Hayakawa, M., Kinoshita, T., & Nio, M. (2007). Revised value of the eighth-order contribution to the electron g − 2. *Physical Review Letters, 99*, 110406-110401–110406-110404.

Bennett, G. W., Bousquet, B., Brown, H. N., Bunce, G., Carey, R. M., Cushman, P., Danby, G. T., Debevec, P. T., Deile, M., Deng, H., & Deninger, W. (2006). Final report of the E821 muon anomalous magnetic moment measurement at BNL. *Physical Review D, 73*, 072003-072001–072003-072041.

Borsanyi, S., Fodor, Z., Guenther, J. N., Hoelbling, C., Katz, S. D., Lellouch, L., Lippert, T., Miura, K., Parato, L., Szabo, K. K., & Stokes, F. (2021). Leading hadronic contribution to the muon magnetic moment from lattice QCD. *Nature, 593*, 51–55.

Brown, H. N., Bunce, G., Carey, R. M., Cushman, P., Danby, G. T., Debevec, P. T., Deile, M., Deng, H., Deninger, W., Dhawan, S. K., & Druzhinin, V. P. (2001). Precise measurement of the positive muon anomalous magnetic moment. *Physical Review Letters, 86*, 2227–2231.

Summary and Conclusions

8

In our Introduction, we began with the thesis, originally proposed by Larry Laudan, that when it comes to evaluating theories, the choices are not just between acceptance or not, but that there is a third possibility, namely, whether a theory should be pursued. This option would allow fledgling theories the opportunity to show their full potential. With the understanding, of course, that there is no guarantee that such potential would, in fact, be achieved. Laudan's claim is that taking judgments of pursuit worthiness into account serves to highlight the underlying rationality in certain key episodes in the history of science.

We proposed that determinations of pursuit worthiness, as opposed to acceptance per se, could be understood, to good effect, as applying to experiments as well—even when there was not an alternative experimental program with a more successful track record. What follows is a catalogue, extracted from our case studies, of what we have described as *The Many Flavors of Pursuit*. As will be seen, there is significant variety here, where the resulting mosaic reveals, through its many interconnections, much of the dynamics of experimental development.

1. Pursuit Worthiness and the Subsequent Experimental Pursuit of the Underlying Promise

As our case studies have shown, there is a difference between deciding that a general sort of experimental venture is pursuit worthy and determining how to put together and conduct an actual experiment to uncover and test the underlying pursuit worthy promise. But the distinction is not nearly so stark as might at first appear. This because an appraisal of pursuit worthiness must consider the favorable prospects for such an experimental test. If there were no such prospects, the venture would not be pursuit worthy.

An example of what we have in mind is afforded by the reaction of experimentalists to the Fifth Force hypothesis. As we have shown, there were many who had both an

© The Author(s), under exclusive license to Springer Nature Switzerland AG 2022
R. Laymon and A. Franklin, *Case Studies in Experimental Physics*,
Synthesis Lectures on Engineering, Science, and Technology,
https://doi.org/10.1007/978-3-031-12608-6_8

interest in pursuing an experimental test, and moreover had already conducted or were planning experiments that could be coopted for such testing. Other interested investigators had worked previously on tests of fundamental laws. And some had even discussed the possibility of a Fifth Force test and had been able to quickly come up with a plan for a simple and workable apparatus along with a preliminary analysis of background and systematic effects. In short, there was both theoretical interest and experimental expertise. Similarly for the Wu experiment where there was Lee and Yang's conceptually simple experimental proposal, the combined experimental expertise of Wu and her colleagues, as well as a known and effective method of Co^{60} polarization.

The developmental path of the Ellis and Wooster 1927 heat experiment was more complicated, and its gestation spanned the period from 1922 to 1927. Ultimately it was realized in 1925 that instead of separately testing for each of the possible contributing factors for the "missing energy" of beta particle production, an existing experimental technology could be utilized to determine the overall energy expense of the beta decay of radium E. In conjunction with this realization, Ellis and Wooster presented an extended argument for the pursuit worthiness of such an experiment which focused on the favorable prospects for the experiment being able to distinguish between the two competing accounts of beta production.

2. Being worthy of pursuit does not require that an experiment be thought likely to prevail against the existing regime of accepted results
This feature of pursuit is most clearly seen in the case of the experimental searches for the Fifth Force. As we have shown, the prevailing attitude among the experimentalists who took up the challenge of mounting an experimental test was that the Fifth Force did not exist. Similarly, the theta-tau puzzle was sufficient to justify experimental pursuit despite the fact that parity was known to be conserved in interactions involving the strong, electromagnetic and gravitational forces, as well as in all of classical physics. Again, the smart money was placed on the experimental tests of parity coming up empty. Nevertheless, Wu and her colleagues pursued and conducted the experiment, and came up with what was for them a positive result.

3. An experimental program may be deemed worthy of continued pursuit despite a succession of negative results
As the cases of neutrinoless double beta decay ($0\nu\beta\beta$) and Supersymmetry show, experimental pursuit may continue despite a dispiriting succession of negative results. In the case of $0\nu\beta\beta$, there were three considerations that justified continued pursuit. First, there was the great reward, that $0\nu\beta\beta$ once discovered could be leveraged to reveal the underlying symmetries that govern leptons. Second, considerable progress had made regarding experimental technique which suggested that an experimental resolution of the existence of $0\nu\beta\beta$ might be at hand. Finally, there was the practical reality that such experimentation

was "the only realistic direct probe for lepton-number violation" (Dolinski et al., 2019, p. 246).

The situation with respect to Supersymmetry is similar with respect to the hoped-for reward, and where its pursuit is buttressed by ongoing theoretical and experimental developments, in particular that while deliberately "simplified models" were used as "approximations", those approximations yielded consistent and experimentally confirmed restrictions on the available real estate that the sparticles could occupy. But unlike the search for $0\nu\beta\beta$ which was the only realistic direct probe for lepton number violation, Supersymmetry accounts do not exhaust the range of possible extensions of the Standard Model.

4. An experimental result may warrant acceptance in some respects while being deemed only pursuit worthy in other respects

Wu's limited conclusion that there had been a "large asymmetry effect" along with the preliminary estimate of the asymmetry coefficient as minus 0.7 was sufficient—and was historically taken to be so—to justify the *acceptance* of the proposition that parity was not conserved in the weak interactions. Moreover, this acceptance was in turn taken to justify the experimental *pursuit* of the consequences of parity non-conservation. Thus, an experimental result may warrant acceptance in some respects while being deemed pursuit worthy in others.

5. A replication of an experimental result is pursuit worthy only if the replication is a better in the sense of being a considered response to a known or suspected confounding factor

An "exact" replication is rarely pursuit worthy—why repeat the same mistakes? Exact replication is warranted only if there exists good reason to think that a relevant causal factor has changed in the interim. More generally, what is required for pursuit worthiness is a proposed modification or variation that shows promise for minimizing or eliminating suspected confounding factors. In short, warranted replication is a species of pursuit worthiness.

Thus, for example, Wilson's 1910 replication of his 1909 experiment incorporated a well-conceived refinement (the initial focusing of the β-rays by means of a second magnetic field) to minimize the confounding effect of a finite aperture. Similarly, Meitner's replication of the Ellis and Wooster 1927 heat experiment took advantage of being able to prepare a RaE source that was significantly purer than that used by Ellis and Wooster, and which also made use of a calorimeter that was more robust and less sensitive to disturbance than that used by Ellis and Wooster.

It must be noted, however, that if the changes in apparatus dramatically change its operation, then the resulting experiment is more naturally described not as a replication but as an *independent* test of the quantity or quantities at issue. For example, Chadwick's 1914 experiment (as it eventually came to be understood) is more naturally described as an independent determination of the beta spectrum as opposed to a replication of Rutherford's photographic determinations of the spectrum. In this case, there was conflict that took some time to sort out.

6. Experiments performed on the experimental apparatus to determine its sensitivity to possible confounding factors are an essential part of every good experiment. In addition to providing evidence in support of either acceptance or further pursuit, such sensitivity testing is often revealing as to specific pursuit worthy avenues of improvement

So, for example, Wilson concluded his 1909 report with an enumeration of four possible confounding factors where the first such factor was the finite size of the apertures used. Here the sensitivity test was to repeat the experiment but without the intervening aperture screens which resulted in absorption curves that made an earlier departure from the straight line of linear absorption. This is a good argument showing the importance of aperture size since it confirmed that increasing aperture size makes things worse. But it does not show, as argued by Hahn and Meitner, that Wilson's apertures were good enough for his claimed results. In response, Wilson introduced an ingenious refinement where the β-rays went through a two-stage focusing process. The experimental payoff was, as Wilson put it, to prove "conclusively" that β-particle velocity decreased as the particles made their way through absorbing media. In sum, Wilson's 1909 sensitivity test justified the pursuit worthiness of a particular type of experimental improvement.

What Wu referred to as "systematic checks" is another example of the importance of testing for the sensitivity to possible confounding factors. Given the large size of the measured asymmetry the results of these tests were sufficient to warrant *acceptance* of the proposition that parity was not conserved. But when it came to a more accurate determination of the asymmetry coefficient, Wu and her colleagues emphasized that "many supplementary experiments" had to be made to determine the various correction factors. The pursuit worthiness of such determinations was thus established by the theoretical importance of the asymmetry coefficient and the fact that there were additional confounding factors such as scatter and "the ever present stray magnetic fields in the immediate vicinity of a polarized β-source" that had to be taken into account (Chirovsky et al., 1980, p. 127).

7. An experiment may warrant either acceptance or further pursuit because of the internal consistency of its interacting components

The Wu experiment is particularly noteworthy in this regard because of the by no means guaranteed coincidence of the disappearance of the β asymmetry and that of the γ anisotropy. This coincidence was without question striking, especially considering that the experiment had to be conducted in two parts with contrary, imposed magnetic fields. It thus served to reinforce the acceptance and pursuit worthiness of the experimental results. The Ellis and Wooster 1927 experiment is similarly noteworthy in this regard when one considers that the complicated procedures employed in measurement and analysis yielded, as an indirect result, a decay period of 5.1 days for radium E which compared favorably with the then accepted decay period of 5.0 days. There was, as in the case of the Wu experiment, no guarantee that there would be this close agreement of decay values.

8. Experimental results may achieve acceptance to the point that even accepted basic principles such as conservation of energy and quantization lose their status as accepted, and such principles and their options are treated as being only pursuit worthy

Depending on the circumstances, an experimental result may be taken as support for the pursuit worthiness of a theory. But the pursuit worthy status so determined may be either a *promotion* of a so far untested proposal or a *demotion* of a theory that had previously been accepted. Thus, in the latter case, the experiment renders the theory in question not refuted but rather demoted to the category of being only pursuit worthy. And this was exactly the effect of the Ellis and Wooster heat experiment which cast doubt not only on the conservation of energy, but also on quantization, and thus served to deny acceptance and changed their status to being only pursuit worthy. Also in contention for pursuit worthiness was Ellis and Wooster's piecemeal combination of classical elements, quantization, as well as Pauli's proposed "neutron", renamed "neutrino". This example is all the more noteworthy because Ellis and Wooster's result was *accepted* as true (especially after Meitner's replication) and not merely as being only pursuit worthy.

9. Experimental programs sometimes take on a life of their own even after their theoretical motivations have been abandoned and continue to be pursuit worthy

Despite what might have been expected, the demise of the Fifth Force did not cause any lack of enthusiasm among experimentalists for developing ever more accurate and expansive tests of Newtonian gravitation and, in particular, of the Weak Equivalence Principle (WEP). If anything, there was an upsurge in interest. In short, the *experimental program* initiated by Thieberger and the Eöt-Wash group (but which had much deeper historical roots) was deemed pursuit worthy even though the initial motivation, the Fifth Force hypothesis, had been deemed not pursuit worthy. There were four reasons for this continued pursuit worthiness. First, even though the Fifth Force was no longer on the scene, the underlying theoretical basis for suspecting that there might be modifications to Newtonian gravitation was still largely intact. Second, there was the accumulated expertise that had been developed as a result of the extensive testing of the Fifth Force hypothesis. Third, there was the continuing fundamental importance of the WEP for the General Theory of Relativity, and, in particular, for attempts to integrate it with the standard model of particle physics. Finally, it was realized that experimental examinations of WEP could be understood and used both as *tests* of WEP and as *methods of discovery* for new forms of non-Newtonian gravitation.

The experimental program of testing the absorption of β-rays by measuring ionization, is similar in that it continued long after it ceased to provide results that were revealing as to the initial production and source of β-rays. As we noted, Rutherford's 1913 tabulation of the principal experimental results regarding absorption does not include Wilson's 1909 result that such absorption was linear and not exponential. Instead, it includes results, all of which indicated an exponential rate of absorption. Similarly, in 1921 Chadwick reported that the consensus was that absorption was either "approximately exponential" or composed of two or three parts, each of which is exponential with a different coefficient. What explains this apparent slight of Wilson's linear result is a distinction that Rutherford made between a *law* of absorption and the *interpretation* of that law—where by *law* he meant an empirical regularity. Accordingly, Wilson's linear results received separate treatment by Rutherford because of their explanatory power. And by 1931, after briefly describing Wilson's "general explanation" of absorption, Rutherford concluded that "[t]he form of the absorption curve appears to therefore *to have small importance* [for interpretation], since being mainly controlled by scattering it is very dependent on the design of the apparatus and the varying extent to which scattered rays are measured" (Rutherford et al., 1930, pp. 414–415, emphasis added). Thus, as shown by this case, initially pursuit worthy avenues of experimental investigation may continue after they lose their theoretical relevance and continue for some time because of their purely empirical value, but eventually (if not taken under the wing of a new theory) become overwhelmed by the confounding effects that have made their presence evident precisely because of that experimental pursuit.

10. An experiment may be deemed not pursuit worthy even though there is no plausible explanation as to why its results are at odds with the results of what are considered to be pursuit worthy experiments
The first two experiments designed to test the Fifth Force hypothesis were by Thieberger and members of the Eöt-Wash group. The Eöt-Wash group used an oscillating torsion pendulum containing two beryllium and two copper test bodies which was located on the side of a hill on the University of Washington campus, that provided the local mass asymmetry needed for an observable Fifth Force effect. Thieberger's experiment was noteworthy because it avoided the complication of having to take account of both gravitational and centrifugal forces since it relied only on the differential Fifth Force effect due to the Palisades cliff in New Jersey. This simplification was possible because Thieberger employed a differential accelerometer in the form of a copper sphere that was partially submerged in water.

The results were not in agreement. Thieberger's results supported the existence of a Fifth Force, whereas the results of the Eöt-Wash group found no evidence for such a force. So, while the two experiments *considered together* had the effect of denying acceptance for both, they were sufficient—given the potential importance of the Fifth Force—to justify further experimental pursuit. But from then on none of the experimental tests agreed with Thieberger's result. Especially significant in this respect were a series of increasingly. more refined floating sphere experiments by Bizzeti and his collaborators.

Considering all this, the consensus was that continuation of experimental searches for the Fifth Force was not pursuit worthy. Nor, was there any attempt to determine what *exactly* had gone wrong in the Thieberger experiment. And Thieberger himself was at a loss to explain what had gone wrong.

The situation is similar, though rather more complex, with respect to the many iterations of the experiments conducted and their associated statistical analyses by Klapdor-Kleingrothaus and his colleagues which were claimed to demonstrate the existence of neutrinoless double beta decay. Here, as in the case of the Thieberger experiment, there was a consistent and mounting cascade of negative experimental results along with pointed criticisms of the statistical techniques employed by Klapdor-Kleingrothaus and his colleagues. So, it was not surprising that their claims ultimately ceased to attract attention, having evidently been determined not to be pursuit worthy—even to the point of not expending any further effort to determine what exactly had gone wrong with either experiment or statistical analysis.

11. In high-energy physics there is a formal criterion for the acceptance of a claimed discovery, namely, that the observed signal must be five standard deviations (σ) above background. But the supporting statistical analysis may be challenged as only being pursuit worthy

As we noted in our discussion of neutrinoless double beta decay ($0\nu\beta\beta$), there is now a consensus in high energy physics that a confidence level of at least 5σ is required as a necessary condition to qualify as a discovery. And assuming all else is in order, that condition becomes sufficient as well. This means that if an experimental result satisfies this condition (in its incarnation as sufficient), then *both* the *result* as it applies to the underlying theory involved, and the *experiment itself,* are to be *accepted.* Thus, in such a case, one need not worry about whether the result and its accompanying experiment are at best only worthy of pursuit.

After an initial series of experiments that did not satisfy the discovery standard, Klapdor-Kleingrothaus and his colleagues claimed to have obtained a result that satisfied a 6σ standard, which was more than required. But on this, other members of the experimental community demurred, not on "the quality of the data" but rather on the soundness of the statistical analysis. And from their point of view—as we briefly reviewed earlier—their statistical objections were made with good reason.

The point to be made about this skirmish between Klapdor-Kleingrothaus and his critics, is that statistical analysis is itself subject to judgements of being accepted, worthy of pursuit or rejection outright. In the case of the claims of Klapdor-Kleingrothaus and his colleagues to have discovered an instance of $0\nu\beta\beta$, there was no acceptance, only an implicit judgment of continued pursuit worthiness. But eventually, as already noted, given the mounting number and sophistication of contrary, experimental results, Klapdor-Kleingrothaus' claims of discovery ceased to attract attention and commentary—and thus, in effect, were not considered as pursuit worthy.

A *corollary* of the 5σ standard for an acceptable discovery, is that satisfaction of at least a 2σ standard is sufficient (assuming all else is in order) to qualify the experiment and its result as worthy of further pursuit. The application of the corollary is evident in the case of the 2σ discrepancy that was found between the calculated and measured values of the g-factor of the muon. That discrepancy generated considerable pursuit in both theory and experiment—to the point of painstakingly moving the experimental apparatus from the Brookhaven National Laboratory on Long Island to Fermilab, in Batavia, Illinois. As a measure of the extent of this ongoing pursuit, we note that as of 2017 more than 100 theoretical physicists had joined the Muon (g − 2) Theoretical Initiative to provide more detailed calculations of the muon g − 2 value, where these efforts indicate that the disagreement between the predictions of the Standard Model and the experimental measurements is now to within around 4.2 standard deviations. In addition, as described earlier, there has been considerable effort on improving the experimental apparatus itself. Thus, pursuit continues on both the experimental and theoretical fronts.

12. The pursuit worthiness of an experiment may sometimes only become apparent in retrospect

What we have in mind here is Chadwick's decision in 1914 to take advantage of Geiger's improved point counter to investigate electron scattering, which was at the top of the list in terms of recalcitrant confounding effects. Such an experimental investigation was clearly pursuit worthy, where the promise of success rested on what was expected to be the more sensitive determinations of the energies associated with beta particle production.

But as we have seen, the project never got off the ground with respect to the problem of scattering. Instead of obtaining data that would pave the way to a better understanding of scattering, what appeared was a surprisingly small discrete beta spectrum that was dwarfed by a continuous range of data values. Chadwick was not pleased because this meant that the scattering problem was going to be very difficult to deal with. It took a while till Chadwick's disappointment changed to satisfaction when it became apparent that his data were better understood as revealing the existence and difference between what became known as the primary and secondary beta spectrum.

Thus, Chadwick's experiment did not deliver on the promise implicit in what was originally thought to be its pursuit worthiness. But in retrospect, and in terms of pursuit worthiness, it delivered most significantly because it had succeeded in separating out the effects of the primary and secondary spectrum. Stated counterfactually, if Chadwick had pursued the experiment because he thought it pursuit worthy with respect to the fine structure of the beta spectrum, he would have been vindicated. What he most definitely did achieve in the real world was a demonstration of the pursuit worthiness of the experimentally revealed distinction between the discrete primary and the continuous secondary spectrum. Given this, the pursuit worthiness *of a replication* was obvious. Thus, in 1922 Chadwick, now with Ellis on board, conducted the replication. Meitner, who well understood the significance of Chadwick's results, oversaw a significantly better replication by her student Pohlmeyer which provided a more differentiated distinction between the secondary spectrum and its nearby confounding effects.

Coda: For the working scientist—theoretician or experimentalist—the question that sets the stage for all else is what is there to do that is pursuit worthy. And even after such a decision is made (or has already been made as a condition for funding or employment), questions of pursuit continue to arise at virtually every stage of constructing a theory or designing and conducting an experiment.

This does not mean that once decided, decisions as to pursuit worthiness cannot be overwritten and replaced with different determinations that reflect ongoing advances in the field of inquiry. Moreover, as amply demonstrated in our catalogue of the many flavors of pursuit, the dynamics of the interaction between decisions as to pursuit worthiness and experimentally motivated responses to such pursuit are wide and varied, with unpredictable and often surprising results. It takes great ingenuity and stamina to negotiate this maze of opportunities and choices. Or so that is what we have been at pains to demonstrate.

References

Chirovsky, L. M., Lee, W. P., Sabbas, A. M., Groves, J. L., & Wu, C. S. (1980). Directional distributions of beta-rays emitted from polarized ^{60}Co nuclei. *Physics Letters, 94B*, 127–130.

Dolinski, M. J., Poon, A. W. P., & Rodejohann, W. (2019). Neutrinoless double-beta decay: Status and prospects. *Annual Review of Nuclear and Particle Science, 69*, 219–251.

Rutherford, E., Chadwick, J., & Ellis, C. D. (1930). *Radiations from radioactive substances*. Cambridge University Press.

Printed in the United States
by Baker & Taylor Publisher Services